Fostering Subsistence Agriculture, Fo
Sub-Saharan A

SUPPORT AFRICA INTERNATIONAL
Studies in sub-Saharan Africa

Vol. 2

PETER LANG
Frankfurt am Main · Berlin · Bern · Bruxelles · New York · Oxford · Wien

Franz-Theo Gottwald
Susan Keino
Timothy Rotimi Fayeye
(eds.)

Fostering Subsistence Agriculture, Food Supplies and Health in Sub-Saharan Africa

PETER LANG
Internationaler Verlag der Wissenschaften

Bibliographic Information published by the Deutsche Nationalbibliothek
The Deutsche Nationalbibliothek lists this publication in the Deutsche Nationalbibliografie; detailed bibliographic data is available in the internet at <http://www.d-nb.de>.

ISSN 1614-001X
ISBN 978-3-631-57380-8

www.peterlang.de

FOREWORD TO THE SERIES

The research findings to be published in the series Studies in sub-Saharan Africa have been promoted by SUPPORT AFRICA INTERNATIONAL and the Munich-based Schweisfurth Foundation.

The articles and contributions are mainly research papers submitted to the annual calls for the Research Award of Sustainable Agriculture by the organizations mentioned above. The findings reflect the present state of research and develoment projects in the respective academic fields at the universities in sub-Saharan Africa. The authors are independent scientists taking responsiblity for their studies published.

Publishing this series is to encourage institutional research and development at sub-Saharan universities. It appeals to members of academic departments, providing them with a platform to voice their concern. Though this series is open to a world-wide readership, it primarily aims at reaching the scientific world in sub-Saharan Africa.

Baldur Ed. Pfeiffer　　　　Franz-Theo Gottwald

TABLE OF CONTENTS

FOREWORD

Franz-Theo Gottwald

Subsistence agriculture is agriculture mainly carried out for survival with few or no crops left for sale. Whereas in rich Western nations, subsistence agriculture has been replaced by industrialised agriculture almost completely, this historic form of land use is still practised in some regions of poor developing countries. For instance, in sub-Saharan Africa, 90 percent of farmers are smallholder. However, nowadays, most of these farmers produce for the market as well, at least to some extent. To these farmers, earning money through agriculture is a minor consideration when deciding what crops to plant and what method to use. Usually, each region relies on traditional crops and methods and the harvest is hardly enough to feed one family

Subsistence agriculture plays an extremely important role for sub-Saharan Africa, and it is closely connected to food supply and health in these regions. Traditional crops and livestock are well adapted to the weather and soil conditions in these regions, and provide valuable nutrients which are vital for a balanced diet. However, subsistence agriculture is highly vulnerable to drought, livestock diseases and pest, and crop loss can have dramatic consequences for the farmers and their families. Moreover, it is threatened by proceeding industrialisation of agriculture in these countries. While in many Western countries farmers try to revert from mass production to small businesses and networks of regional supply and marketing, in developing countries, more and more farmers replace traditional crops and methods by new, expensive techniques, following the misbelief that this could ensure their existence. But quite the contrary is the case; relying on new technologies often leads to high debts due to expensive seeds and machines, and since many farmers do not own the land they depend on or own only very small parts of land, capital investment is not advantageous to them. Furthermore, food supply by subsistence farming is put at risk by HIV/Aids. In some regions, the parental generation has almost died out, leaving behind orphans and the grandparents, which are hardly capable of caring for themselves.

Since the chances and problems regarding subsistence agriculture are relatively complex, it is particularly important not to focus on only one or two aspects, but find practical and sustainable solutions for all factors involved. Introducing new concepts, human health and diseases have to be considered as well as malnutrition and underfeeding, the role of women in agriculture, the attitude of farmers, market conditions, infrastructure, plant and animal health, weather and soil conditions and the given social and political structures in the countries of concern.

Although subsistence agriculture is often based on organic production – simply due to the lack of money to buy fertilisers, pesticides or genetically modified seeds - it is not always sustainable. For instance, in rainforest regions harmful techniques such as slash and burn are relatively common. Deforestation and illegal hunting of endangered species indeed is the less popular side of agriculture in Africa. However, these problems cannot be solved merely by mechanisation and modernisation. Industrialisation of the agricultural sector is rather likely to be even more harmful in respect to environment and wildlife

Subsistence agriculture also can have an enormous impact on rural areas and therefore make a major contribution to the sustainable development within a region. First, it evolved "from the region for the region", which means that it is perfectly adjusted to the local conditions and meets the needs within a certain area. Second, it preserves work opportunities and adds new jobs, building up social networks, which are characterised by mutual help and support. Social structures are perpetuated which is particularly important regarding the tremendous societal reorganisation due to migration and HIV/Aids. And third, subsistence agriculture is sustainable, because it takes responsibility for the environment and for future generations.

In industrialised countries, agriculture is characterised by exploitation and a consumptive and wasting manner of land use. Huge pig or beef farms with 80. 000 animals or more are exhausting natural resources, and when soil, air and water are polluted, farmers simply move to the next location. This irresponsible and ignorant way of land use is inconceivable in subsistence farming, where the vulnerability of nature is omnipresent. Land and natural resources are not only regarded as monetary factors, but as an essential basis of life. The traditional thinking is still predominating, that the land can be used for cultivating crops and rearing livestock, but the natural resources have to be conserved in order to ensure the survival in the long term.

The Schweisfurth-Foundation is particularly engaged in supporting rural development. Food production, processing and marketing should be based on the "ecology of short distances". This principle is the impetus for an economically and ecologically reasonable and salutary development of rural regions. In the rich Western countries, the centralised and globalised economic structures have led to economic, social and cultural losses which have serious consequences for rural regions. Therefore, we initiate and support projects which aim at strengthening rural culture and development and enhancing networks between farmers, processors and suppliers. The "ecology of short distances" is based on multi-functional and regional creation of value. Besides the production of food and agricultural products, it contains several other fields such as renewable energies and primary products, health, education, tourism, environmental protection, waste management and technological development.

The following studies are part of the scientific competition of the *Research Award in sub-Sahara Africa*, which is annually organised by Support Africa International e.V. and the Schweisfurth-Foundation. Since 2000, young researchers from sub-Saharan Africa are invited to take part in this competition in order to win the prize money and get the opportunity to present and publish their research to a wider public. These articles summarise the most urgent problems with respect to subsistence agriculture, nutrition and health. By presenting well-founded scientific knowledge and showing a deep understanding of the topic as well as personal commitment, innovative solutions for problems concerning agriculture, sustainability, health and food supply in sub-Saharan Africa are pointed out.

I am grateful to Monika Weiler-Helbach and Dietrich Schwarz for their editorial support.

STRATEGIES FOR FOSTERING SUBSISTENCE AGRICULTURE

Abiodun Ade Adeloye

Nigeria has a population of about 120 million, and 70 percent of this lives in the rural communities and are engaged in subsistence farming as their means of livelihood. Poverty is pervasive and the guidelines for the implementation of the Poverty Alleviation Programme (PAP) of the Federal Government revealed that 60 percent of Nigerians live below the poverty line and only 50 percent of the population has access to safe water. About 38 percent of the population does not have access to primary health care while most Nigerians consume less than one-third of the required proteins and vitamins due to low purchasing power. Poverty reduction strategies have been declared operational at the capital city and are supposed to be implemented. However, reports from the rural communities show that this measure is yet to reach the people.

1. Economic and political framework

Nigeria is a constitutional authority with a democratic government. It has in the south an opening to the sea and some parts of the country has a rugged terrain. Rainfall is heavy in the south and decreases to the north of the country. Of the total land area of 93.3 million hectares only about 34 million hectares are cultivated. The total cultivable land area is estimated at 71.2 million hectares (Nigeria 1975). Increased competition on the most fertile land takes place mainly between agriculture on one hand and settlements and other land developments on the other. Nigeria maintains diplomatic relations with more than 80 countries in the world. Nigeria is a member of the United Nations (UN), Commonwealth of Nations and the Organisation of African Unity (OAU) that has been reconstituted as African Union (AU) in 2002 and ECOWAS. The economy of Nigeria depends mainly on the crude petroleum, agriculture, wholesale and retail trade, finance and insurance as well as producers of government services. The tables below indicate the development of some key economic ratios, the distribution share in the Gross Domestic Product (GDP), stakeholders in economic and rural developments. Nigeria is characterised by a low degree of decentralisation. Key administration and political decision making functions are focused on central institutions. With the exception of the rural towns, the rural economy is highly concentrated on and around agriculture and homesteads, either providing products and services for agricultural and domestic

Stakeholders in rural development

Stakeholder	Level	Main interests contribution
1. Government and Administration		
ministry of agriculture	central, local district	support to farmer (training & advice, distribution of subsidies, marketing) divestitutre of state enterprises
various government institutions, farmer support services	local district	veterinary services, provision of agricultural implements, lending of machinery, provision of mechanised services (partly subsid.)
ministry of local government	contral district	development planning, land allocation administration, collection of ground rent and taxes
2. Civil society		
traditonal chiefs	local	land allocation control of development activities
co-operatives and associations	local (production), central district (services)	sharing of means of production, profit oriented production of crops and
NGOs	all	support to specific target groups
3. Private sector		
small scale farmers subsistence and commercial	local	support to specific target groups, production of crops and livestock (products and
medium scale commercial farmers, few	local	profit-oriented production of crops and livestock
local private entrepreneurs (farmers support services)	local district	veterinaries, lending of machinery, mechanised services and agricultural (profit oriented) basic food processing , few advanced)
South Africa private entrepreneurs (farmer support services)	foreign cross border business	lending of machinery, provision of specialised services and agricultural implements (high investment costs like combined harvesters of agricultural planes)

Table. 1 Stakeholders in rural development (based on Kingdom of Lesotho, 2002a). It shows the major stakeholders in rural development and their main level(s) of interaction and their main interest or contributions to rural development as indicated on page 12.

activities or using and processing agricultural products. Agriculture is dominated by subsistence farming on one hand where only households classified as "wealthy" or "better off " do not suffer from at least 1-2 months of food shortage per year. On the other hand livestock production and grazing are predominant whereever the land cannot be used for farming. Overstocking and unproductive cattle are common features due to cultural reasons. Intensive crop and livestock production or processing of agricultural products for domestic and external markets rarely take place. In recent years, tourism has been seen as a potential development priority area by the government.

2. Core problem and major constraints

The rural economy is mainly focused on agriculture, whereas agricultural productivity (crops and livestock) is generally low due to difficult edaphic and climatic condition, but also to poor land management, land tenure and land use, rights arrangements are insecure, land is unequally distributed, land parcels are fragmented. These factors, amongst others, hamper needed investments in the increase or at least maintenance of agricultural productivity.

Potentially good land for agricultural purposes is a scarce resource, concentrated in areas with high potential also for non-agricultural development. The need for land for settlements and businesses in those areas on one hand and the low degree of application of intensive farming methods on the other increase the pressure on land. At the same time, labour force required for both intensive agriculture and non-agricultural activities is lacking due to a high degree of migration from rural to urban areas. This labour migration is again induced by a lack of employment opportunities in the rural areas. This problem area is aggravated by the absence of rural finance facilities. The motivation to invest in the intensification and thus modernisation of the agricultural sector and therewith the ignition of a process of developing related economic activities is very low.

Drought is a serious ecological problem. The traditional explanation for the droughts particularly in the northern part of Nigeria is that it is caused by the army of worms. The belief is that when the rain clouds gather, the army worms lie on their backs with their feet pointing to the clouds and use this to disperse the rain clouds! However, there is awareness of the loss of traditional farmlands to city prospectors who buy the lands and engage in illegal mining to prospect for precious stones. These activities bring up large quantities of subsoil to the surface and leave several deep trenches, thus destroying the farmlands. There is therefore a demand for new farm lands which are obtained by cutting down the forest. The attendant forest destruction is the cause of drought and unchecked forest clearing.

A significant proportion of the rural dwellers are illiterate, being unable to read and write. The common language spoken is the dialect and has its drawbacks in providing an effective communication medium. Most printed information on agricultural activities is in English and very rarely in major dialect i.e. *Yoruba, Hausa and Igbo,* even though such information is not readily available. It is almost elitist to be seen reading one of the leaflets containing information on agricultural activities. Besides, the rural communities have a rich oral tradition by which information and knowledge is preserved and transmitted. The limitations of such a repository mean that some vital knowledge is inevitably lost, or inaccurately stored.

3. Food crisis in Nigeria and government attempts

The Nigerian economy has been going through some difficult periods in the last decade or so. Perhaps one of the most noticeable problems is the emergence of an era of shortage. In the 1970s and up to the 1990s and particularly in the 2000s, food shortages have reached a crisis stage and has remained that way ever since. A lot of factors have been identified to explain the causes of the phenomenon. These include the emergence of the oil boom, which led to the neglect of the agricultural sector by the authorities, the drift of farmers to urban centres from the rural areas as a result of the adverse terms of trade against agriculture, and the stagnant technology in peasant agriculture among others. Compelled by the growing scarcity of food available in Nigeria, the governments have been making efforts to increase food production. Such efforts vary, although their effectiveness is not easy to determine. This is mainly because the same traditional method of supplying farm inputs to farmers is still being applied. There is need to have a consistent and realistic policy aimed at increasing available food supply in Nigeria to avoid the present structural contradiction of surplus capita, land and labour coexisting with increasing food shortage.

There are three possible policy directions to increasing available food supply in Nigeria. The first is one which relies entirely on private food producers while providing necessary incentives and facilities. The second is a combination of private and public food production. The third is one which relies entirely on public corporation and agencies.

The agricultural economy of Nigeria is dominated by small-holder farm households, a situation in which approximately 60-80% of the total population live in rural areas where most of the agricultural activities take place. Approximately 95% of the farms are small holdings, at most 4 hectares in size, with family labour furnishing the bulk of the labour force (Adesimi et al, 1985). Some of the problems of small holders (subsistence farmers) are economic, some are physical, other are cultural while some relate to problems which have their roots in sub-optimal enterprise combination (crop mix) or in socio-cultural altitudes, and the community norms which require farmers to grow sufficient food to feed their families without resorting to the market for shortfalls (Ukagha, 2003). This latter factor implies that subsistence requirement is one of the major elements in food production decisions of the small farmer. This should be recognised explicitly in any planning model that seeks to capture the goals and constraints of the small-farm in his attempt to organise his resources for farm production.

Government intervention is, however, a necessity in agriculture. This is just being recognised. In fact, the current emphasis on agriculture in Nigeria's economic growth is comparatively recent. The 1950s witnessed an era of import substitution, because of the pessimism of export promotion and utmost reliance on industrialisation as the dynamic element of economic growth (Fd. Govt., 1962). In the past, agricultural development policy had no definite focus on the production of food crops but on export crops like cocoa, rubbers, etc., which were the major sources of government revenue and foreign exchange earnings (Eicher, 1967). The 1970s witnessed an era of "oil boom", when petroleum dramatically took over the major function of foreign exchange earning from cash crops. However, between 1970 and 1974, the Nigerian government began to give some attention to the production of food crops in the country. Certain socio-economic events of that period such as the payment of arrears of "Udoji" salary award, the just ended civil war and so on led to great a scramble for food and a huge amount of money was spent on for imports.

A number of food production programmes together with a corresponding increase in the financial allocation to food production were made. Among the projects embarked upon include the national seed multiplication, the Nigeria Credit Bank among others. Unfortunately, a lot of the programmers initiated by the various governments to step up production of food crops focused specifically on large-scale farms or gigantic agricultural projects. For instance, the Western Region Farm Settlement Scheme of the 1960's (a highly expensive scheme) aimed at providing jobs for young school leavers and increasing food production, The National Accelerated Food Production (NAFPP) design to bring improved practices such as high yielding seed varieties credit and management to the farmers through co-ordinated package approach; the Operation Feed the

Nation (OFN) programme designed to mobilise the entire populace to participate in food production, the River Basin Development Authorities designed to exploit water resources for the production of food; and the Green Revolution aimed at making the country self sufficient in food production. An evaluation of the effectiveness of these various strategies in achieving their goals by Okuneye (1985) showed a poor level of performance. It is observed that up to-date no well directed and vigorous executed government agricultural policy or programme has been implemented to significantly affect the fortune of the Nigerian small farmer.

This view is supported by series of studies (Adesimi, 1976; Idachaba, 1980; Adeniyi, 1988). Significant contributions to increase food production cannot be made in either government or public corporation plantations, which are periodically plagued by labour disputes and where renumeration of the workers is based on standardised wages and other benefits unrelated to the profits being made by the enterprise or the productivity of the workers. Hence, our reliance on food production from public corporation like the River Basin Authorities should be looked upon as a panicky measure with a good short-run effect. A more lasting long run solution should be the one that seeks to reduce direct government participation in food production in favour of private entrepreneurs. It is a well-documented fact that the unit cost of food production by public agencies is much higher than that of private producers. It, therefore, makes more economic sense to reallocate resources in favour of these low-cost producers.

4. Strategic Approaches:

The main strategy to address the core problem of subsistence agriculture should be the one that leads and supports agriculture and promote rural development through:

i. Access to sufficient, safe and nutritious food.
ii. Eliminate skewed participation and inequity in the sector.
iii. Maximising growth, employment and income in agriculture.
iv. Enhancing the suitable management of natural agricultural resources and ecological systems.
v. Ensuring knowledge and information management.
   Priority programmes and actions have been identified (SA, 2003).
vi. Implementing a broad-based safety and security strategy for good working and social stability, trust and confidence.
vii. Fostering a shared vision for agricultural good governance and social partnerships.

viii. Fast-tracking a programme of land redistribution for agricultural development and processes of empowerment for targeted groups.
ix. Transforming agricultural research and transferring technology and education to be more responsive to markets.
x. Redefining the mandate of agricultural marketing and international trade in the post-control board era against great global competition and demands for market access, infrastructure and information.
xi. Building credible agricultural, statistical and economic analysis systems that will be accessible to all farmers and enterprises.
xii. Building an agricultural risk-management system to ensure that agricultural enterprises will recover, grow and develop after natural disasters.
xiii. Developing an effective integrated risk management system for plant and animal-health systems, price and income systems and natural disasters.
xiv. Establishing an integrated rural financial service systems infrastructure, irrigation, electricity, telecommunications.
xv. Targeting investments in rural areas to provide livelihoods, transportation, training and skills development.
xvi. Establishing an agricultural co-operation programme for Africa to spearhead the role of agriculture in the New partnership for Africa's Development (NEPAD).
xvii. Lowering the overall cost of production.

The following outcomes are expected from the successful pursuit of these strategic objectives:
i. Increased wealth creation in agriculture and rural areas.
ii. Increased suitable employment in agriculture.
iii. Increased income and foreign exchange earnings.
iv. Reduced poverty and inequalities in land and enterprise ownership.
v. Improved farming efficiency.
vi. Improved national and household food security.
vii. Stable and safe rural communities reduced levels of crime and violence and sustained rural development.
viii. Improved investor confidence and greater domestic and foreign investment in agricultural activities and rural areas.
ix. Pride and dignity in agriculture as an occupation and sector.

There are many factors affecting the development of subsistence agriculture in Nigeria. These include the land use systems, research infrastructure, extension service and socio-economic environment. Others are the physical conditions (weather, soil).

4.1 Land tenure: In a situation where farmers have no permanent title to land, there is reluctance to adopt improved land management and conservation practices. The less secure the land ownership, the greater the tendency for the farmers to exploit it using non-sustainable practices. Credit is often unavailable to tenant farmers and consequently such farmers are unable to invest in new seed varieties, agrochemicals and new technologies.

The main strategy to solving the problem is by comprehensive land reform based on an approach that simultaneously interlinks the federal, state and local levels. A holistic approach dealing simultaneously with land policy and legislation land administration and information systems and land related planning is recommended (Selebalo and Effler, 2002). The structure of components recommended is as follows:

Table 2: Structure of components of the land reform approach

Components	Output
Land policy and legaislative reform	Land users and institutions dealing with land are guided by a consistent national land policy and related legislation.
Land administration	The land administrative system has the compassable land-related information and to regulate allocation and use of land.
Land information system	A national system for acquiring processing, storing and distributing land related information is operational.
Land use planning	The capacity of population and institution in planning of land use is improved and contributed to a sustainable of land.

This approach provides a framework of policies and actions designed to establish and thereafter ensure the maintenance of a system of land tenure and land management that will facilitate:

i All citizens having the opportunity to occupy and use land.

ii The efficient, effective and economically viable operation of land market.

iii The appropriate regulatory arrangements for the efficient, sustainable and equitable occupation and use of land.

This approach is expected to stop illegal settlement, encroaching into agricultural land.

Well operated land markets will also include regulating and securing short-term traditional leasing and share cropping practices leading to more fields being used productively. The concept of treating land as a commodity is foreign to most African traditions, yet the economic environment and globalisation dictates that these beliefs be reassessed. The reality is that there are already extra-legal transactions where agricultural land is subdivided and sold for residential developments especially in peri-urban areas. It is expected that improved land tenure security, combined with improved access to moral based services will lead to more efficient kind use practices. This, together with improved land management and the design and implementation of land use and development of local authorities will have a strong positive impact on the motivation and ability of people in the rural areas to invest in agriculture and agro-linked businesses, however small these investments may be under the given conditions.

Land reform, however, should not be undertaken in isolation by government excluding other stakeholders. In agreement with Selebalo and Effler (2002), land reform be a consultative process with participation of all levels of communities, NGOs, local government structure, the private sector and the community. It is common for governments to overlook this important requirement of a participatory land reform process; therefore it is essential to develop strategies to develop to strategies to address the deliberate or accidental oversight on the part of government. This involves creating capacity in community based organisations and Non-Governmental regional and international networks that should avail support in terms of relevant materials, studies and technical supports.

Regionally, there are a number of land reform programmes being implemented in Southern African Development Community (SADC) and beyond such as Botswana, Lesotho, Malawi, Tanzania, South Africa, Swaziland, Uganda and Zimbabwe. There are lessons and experiences to be shared amongst these countries such as the need for policy adoption before development of legislation methods in the consultative processes, open dialogue between government and stakeholders and the role of development partners as facilitators of the process not imposing imported perceptions.

However, it would be futile to assume that a workable solution in one of these countries will automatically function in another country. There are subtle diversities amongst these countries that require careful analysis before making conclusions.

The sustainability of the land reform process is also questionable. The process could come-up with the best polices and laws, but if the authority lacks institutional capacity and the will to implement the proposals the effects will come to nothing. The previous and current land legislation have failed, because they were not known beyond the capital cities, therefore sensitisation should be of high priority. The procedures for advocacy and follow-up should be developed. A ministry dedicated to land issues would prioritise the implementation of the land reform and seek the necessary resources of higher levels. Continuity of the process will be ensured through the establishment land reform commission, which would oversee the implementation and make amendments as and when necessary.

4.2 Research: The adoption of technologies by farmers depends on many factors and includes the research, and socio-economic extension services, and socio-economic environment. Others are the physical conditions (weather, soils) and land use systems. If only one of the variables is unfavourable to the receiving client or end-users, the rates of adoption will certainly not be satisfactory. In the past studies carried out to find out why farmers face problems in adopting recommended production practices identified the following major obstacles (Rudat et al, 1993).

i. Insufficient qualified and motivated extension services.
ii. Solutions are no relevant (research agenda).
iii. None or limited credit facilities.
iv. Unreliable market system for input and outputs.
v. Traditional and cautious behaviours of farmers.

These observations (except for the last one) indicate the weakness of the system as a whole and not necessarily that of individuals. Research whose recommendation will be accepted by farmers should have the following objectives:

i. Get a clearer picture of farmers problems and the magnitude thereof.
ii. Be able to rank problems according to their importance to the farmers.
iii. Better define a research agenda at the various levels.
iv. Be able to do good field experimentation in farmer's field together with the farmers and the extension services.
v. Be able to develops adequate communication between the extension services and farmers.

To achieve these objectives a common policy on research and extension must be developed. Research must be farmer oriented, and research must learn to co-operate with farmers and additional front-line staff. Research funds must be reallocated according to this approach. Logistics play a vital role in this respect. Reliable services must be available. The main objective of this approach is to make available to farmers innovations, which are compatible with available resources and the objectives of the farming households. This requires not only planning procedures, but also a thorough understanding of the socio-economics and agro-ecological settings under which farm households live and operate and requires also their participation in the development process.

4.3 Many Non-Government Organisations are known to play vital role in agricultural development (Rudat et al, 1993; Talabi et al, 2002). NGOs have well-trained and motivated staff and are sufficiently equipped to work in the field and with the rural dwellers. Adoption of innovations at farmer level is one of the highest, because they work according to well-defined professional and religious objectives. Fantsuam Foundation is an example of NGO in Nigeria whose efforts in empowering rural women, providing them with micro-credit and access to information and communication technologies is based on the philosophy that the client usually has the knowledge and skills that can form the basis of self-improvement strategy, and therefore begin always with what knowledge is available within the communities, work with these and build on them. Fantsuam Foundation has photographic documents of the communities, and these come handy and effective in explaining that one of the causes of the drought is the unchecked forest clearing that has been taking place (Talabi et al, 2002). The use of audio-visual presentation of ecologically unsustainable farming practices has been reported effective in starting a debate on what options there are if future droughts are to be checked (Talabi, 2002). The use of video clips builds on the peoples' strong oral culture, describing the familiar local agricultural landscape, and relating it to an immediate social concern. It presents facts in a culturally acceptable and accessible format, using local languages and familiar faces for narrations. Its timeliness also means it is presented at public fora at a time when the issue is topical and the scientific basis of the drought is easier to explain because of the experience is still fresh in peoples' minds. Fantsuam Foundation uses a broad range of facilities as audiotapes, video clips, community theatres, and world space radio receivers.

Community theatre group (mainly youths and women) will be engaged in producing education and entertainment, which can be captured on videos and shown to other audiences. Some rural communities have invested their disposable incomes in battery operated television. VCDS are becoming cheaper and these could be used in community groups to show locally produced CD.

The provision of community radio programmes can be an effective and powerful way of communicating with the rural communities, especially using the solar powered suitcase radios. The community and interest focus of community radios make them effective in reaching different targets within the communities and in the different languages spoken by such people. It provides an entertainment information and education forum which is easily adapted for topical issues that are of interests and concern for its communities, but that may not be of interest to larger urban audiences.

Unfortunately, the licensing procedure for rural radios in Nigeria is very expensive and cumbersome for rural communities. The requirement of prior approval of such licenses makes the process even more intimidating for the poor communities. Efforts should therefore be made at ministerial levels to convince communication policy makers in Nigeria of the need to make radio licenses more accessible and affordable to rural communities that are poorly served by the urban based state radio.

4.4 Role of women in agriculture

Traditionally, Nigerian women and their counterparts in other African countries are mainly homebound, nurturing the family and in some cases assisting their men folk on the farms, in fishing and in some cases in trading. Their role tended to be supportive in nature. From early childhood, the female is brought up to be seen generally, but not to be heard. The male is encouraged to be aggressive and assertive. The female started as somebody's daughter or sister and invariably ended up somebody's wife or one of his women. Early educational opportunities discriminated against the female as fathers preferred to educate the male.

Other socio-cultural and demographic factors also influence female participation in agriculture. These factors include societal norms and attitude, age distributions of the labour force, the level and rate of female education, religions and the structural pattern of development. Statistics show that more men are educated and trained in Nigeria than women as indicated by 16% female literacy compared with 31% male literacy (FOS, 1981).

Women are the key producers in rural economy. Women headed households are dominating the set up in many villages, characterised by two major groups; widows and divorced women; and women whose husband are migrant labourers (Selebalo and Effler, 2002). However, in spite of the key role-played by women in subsistence agriculture, they are not entitled to take decision on land issues. There is no doubt that the issue of women's poor access to land rights based on customary tenure arrangement is a major constraint to subsistence agriculture.

About 60% of Nigeria rural populace are women and more than half of them provide 70% of thee rural abour force (Igben, 1980). Regardless of the size and level of integration of a household, women are found to be involved in agriculture (Ala, 2000). The nature of their involvement varies greatly according to religion, ethnic group and wealth. Adeloye (1998) observed that women has more liberty and could engage in farming activities as a full time or adjunct to the usual work, and such activities are at subsistence level. The involvement of the women in large scale agriculture in Nigeria may take sometime to materialise. The main reason is that gender equity is yet to find a place in the social custom in the country. The man takes the decision and even when he is lazy and unproductive, he is still recognised as the one who controls the income. Encouraging the women through NGOs is advocated as one major strategy for fostering subsistence agriculture. For example, the Fantsuam Foundation, an NGO on Nigeria, work with women in rural communities, providing them with access to microfinance services and ICT facilities. Talabi et al (2002) reported that women use most of their microfinance loans for agricultural activities such as farm inputs, grain buying and selling etc.

4.5 Extension services:

The work of extension service is to help farmers to improve their farming practices. Good extension service try to response to community needs, rather than just carry out government policies. But often there is conflict between the needs of the community and the policy of the government. Then extension officers as forced to choose between the community and the government. If both parties want the same thing, then a lot of work can be done by extension officers.

Extension officers are often frustrated when they advise farmers on some farming practices which are too expensive for subsistence farmers, so often the advise is only useful to rich farmers.

The training given to extension officers seems to be deficient in some practical education, e.g. knowing how to measure the fields, knowing how to make fences, knowing about diseases and how to give medicine to animals. Many extension officers have never actually done them. In their training, the extension officers are not taught to work with women. They are taught that men are the decision makers. But in fact most men are migrant workers in the towns. Women do not only hoe the fields and keep the crops, they also decide when and what to plant.

It is suggested that good extension officers should have the knowledge of how to work with those groups in the community, which have the greatest needs. They should be able to talk to farmers and understand that problems cannot always be solved in the way they have been taught, and so they look for different ways.

If people do not have land, they help people to do things at their homesteads, by working with poultry groups, gardening group and groups growing fruit trees for example. This kind of organisation is the best way for extension officers to serve the community.

Sometime this means that extension officers have to go against their training and against government policy, but it is sometime only in this means that improvement and change can happen. When extension officers become community organisers, their work can really help the people.

Summary

Fostering subsistence agriculture in Nigeria demands a thorough knowledge of its potentials and limitations in Nigeria. One of the crucial points is probably the problem of finding appropriate technical, cultural and socio-economic solutions for development and building up a solid, functional and responsive infrastructure for development of approaches to subsistence agriculture is needed to foster the expansion of food production to feed the growing population in a sustainable manner enforced is as follow:

i. Land reforms should not be undertaken in isolation by government excluding other stakeholders. It should be a consultative process with participation of all levels of communities, NGOs, local government structure, communities. This approach can be sustained through the establishment of a land reform commission, which would oversee the implementation and make amendments as and when necessary. Under this reform, laws and practices discriminating against women should be abolished. However, government should note that the removal of certain clauses would not change cultural and social behaviours. Therefore, there should be strong sensitisation and training campaigns.

ii Research and extension services must include the farmers as equal partners if subsistence agriculture is to be successful and sustained.

iii. Technology transfer: Effective research and development in agriculture involves vertical and horizontal transfers of information vertical transfer, involve the flow of information from the farmers to researcher and back to farmers through extension services. The research priorities have to be set with clear appreciation of farmers needs and constraint. The horizontal transfers involve exchange of information about experiences and solutions to common problems between regions or groups. This can be achieve through conferences, workshop, scientific exchange and information media. That transfers of new technologies from advanced research institutions outside Africa sub-region could be involved and the assessment of the applicability to local agro-ecological conditions must be considered.

International organisation could help to facilitate information net working among African scientists and between African and non-African scientists through:

a. Animal scientific meeting on fostering subsistence agriculture.
b. Provision of newsletters, journals periodicals, books, etc. to African scientists.
c. Provision of facilities for publication and bi-linqual translation.
d. Compilation of electronic databases on relevant information and provision of exchange opportunities among African Scientists.
e. Encouragement of women through NGOs, for providing them with microfinance services.

References

Adeloye, A. A. 1998. *The Nigeria Small Ruminant Species,* pp 68-69. Corporate Office Max. Ilorin Nigeria..

Adeniyi, W. R. 1988. "Government Programmes, Agriculture and The Small Farmer- A Review." *The Agric. Educator* 2:84-94.

Adesimi, A. A., D. Planut, Y. I. Fabiyi, and O. Ajobo. 1985. Proposal for the Establishment of Farm Management Department, Faculty of Agriculture. Obafemi Awolowo University, Ile-Ife, Nigeria.

Ala, L. A. 2000. "An Economic Analysis of Livestock Production by Women Farmers in Sokoto State." *Sokoto J. Agric. and Envoi* 2.2.

Eicher, C. K. 1967. "The Dynamics of Long-term Agricultural Development in Nigeria" *J. Farm Economics* 49(5):1158-1170. Lancaster, Federal Government of Nigeria. 1962. Nigeria National.

Development Plan. 1962 – 1968. Lagos. Federal Ministry of Economic Development.

FOS, 1981. Federal Office of Statistics. Lagos, Nigeria. F. S. Idachaba, 1980. Concepts and Strategies Integrated.

Rural Development: Lessons from Nigeria. Department of Agricultural Economic, University of Ibadan, Nigeria.

Igben, M.S. 1980. "Socio-economic Activities of Women in Selected Nigeria Rural Area/communities: Implication for Integrated Rural Development." A Report Submitted to CANCER University of Benin.

Okuneye, P. A. 1985. *The Effectiveness of Collective Farming Programmes in Africa.* NISER, Ibadan, Jimmy

Rudat, H., J. Van Barger, and E. Frey. 1993. "Some Problems Related to Adoption of Technology by Farm Household Sustaining Soils. Sustaining Soil Productivity in Intensive African Agriculture." Seminar Proceeding, Accra, Ghana Nov. 1993, pp 67 - 70.

SA. 2003. *South Africa Yearbook 2003/2004.* pp 65-67.

Selebalo, Q. C. and D. Effler. 2002. "Fostering Rural Development through Agriculture Based Enterprises and Services." Int. Workshop. 20-22. Nov. 2002. QTZ – Haus, Berlin.

Talabi, Y., C. Kazanka, J. Dada. 2002. "Army Worms, Drought and Micro-Finance a Nigeria Experience." Int. Workshop. 20-22. Nov. 2002, QTZ-Haus. Berlin. pp 2-4.

Ukagha, E. 2003. "Socio-economic Constraints to Livestock Production in Nigeria." *Proc. 8th Ann. Conf. Sci. Ass. of Nig.* Sept. 16-18. Fut, Minna, Nigeria.

STRATEGIES FOR FOSTERING SUBSISTENCE AGRICULTURE IN SUB-SAHARAN AFRICA

Augustine O. Angba

Abstract

The need to increase food production to meet the food and nutritional needs of the ever growing population not only at national levels but also at individual and household levels has necessitated this attempt to unveil the strategies of fostering subsistence agriculture in sub-Saharan Africa. To achieve this, the concept of subsistence agriculture, its place in a national economy, the models and strategies of subsistence agriculture were examined. The concept of subsistence agriculture is not limited and restrictive in scope as perceived by many. Its intensity and dimensions are revealed in the role it plays not only in providing good food with high nutritional value, immediately available to the rural people but also in creating the basis and formation stages of commercial agriculture. The farmers, governments, extension and other instruments involved in subsistence agriculture are the focal point of change. In conclusion, renewed efforts in promoting technological change in subsistence food production along with commercial agriculture while ensuring adequate household food consumption both in quality and quantity will form the basis for a subsistence agriculture and economic development.

Introduction

Agriculture is the backbone of the economies of most African countries. It is often perceived by governments as a strategic sector that should, to the extent possible, result in self-sufficiency in staples and in food security. Agriculture must meet the requirements of the ever-growing population by increasing production and improving productivity in an environmentally suitable and sustainable manner. It must also diversify and improve the quality of farm produce as incomes rise. Progress in agriculture is closely linked to several and often inter-related factors such as infrastructure, research, extension, farm inputs, rural credits, price policies and marketing etc.

Attaining a reasonable level of progress through sustainable production is a prerequisite for improving the standard of living of the people and consequently reducing poverty and mortality rate in sub-Saharan Africa. The sub-Saharan

Africa is predominantly a rural region in which agriculture plays a fundamental role in people's work and subsistence. Redding (1998) believes that there is not anything called subsistence agriculture in Africa today, except on a very small scale and in limited regions. Subsistence agriculture is said to exist where production and markets have been disrupted by war and disease. Commercialisation is conceived as an alternative to subsistence.

Commercialisation of agriculture is occurring rapidly in some African countries while in others, there is stagnation of even a return of subsistence agriculture on a large scale. The latter in the views of Joachim and Eilean (1994) occurred because of economic restructuring and the shift from planned to market-oriented economies that have resulted in increased income and employment risks. About 440 million farmers in developing countries still practice mainly subsistence agriculture on much of their land forgoing the potential benefits from domestic and international trade. The subsistence crops cover more than 50 percent of land resources in the majority of low income countries, yet land constraints, ecological problems and rapid urbanisation call for change (Joachim and Eilean, 1994).

Historically, most farmers in the opinion of Wikipedia encyclopaedia were engaged in different farm practices for subsistence and this is still the case in the poorest developing nations, where over one billion people live on under one dollar per day and two more billion live on under five dollars a day. To half of the human population on earth, money is a minor consideration in how agriculture works, what crops are planted, and what tools and methods are used. Each region has a few traditional crops, and methods to grow those that amount to gardening on a large enough scale to feed one family or extended family.

Subsistence Agriculture is often seen basically from the negative point as displaying low responsiveness to policies and therefore, it is difficult to control and direct. It is also perceived by many to be synonymous with backwardness and inefficiency by holding down economic growth and economic performance. These reasons according to Bruntrup and Heidhues (2002) is the cause why many governments have tried in futility to eliminate subsistence agriculture.

In view of the inevitable role subsistence agriculture plays in rural African agriculture, there is a need to determine the strategies that can foster subsistence agriculture in sub-Saharan Africa. To achieve this, attempts were made to examine the concept of subsistence agriculture, the role it plays in economic development, the models and problems of subsistence agriculture.

The concept of subsistence agriculture

The term subsistence agriculture is said to be synonymous with such other concepts as traditional agriculture, small scale, peasant, low income, resource poor, low inputs or low technology farming. Many of these concepts, in the views of Bruntrup and Heidhue (2002), are also used in non-economic disciplines with very different meanings. Thus it becomes difficult to give a generally accepted definition of subsistence agriculture.

Subsistence means supplying needs from the local environment. There are practically no truly subsistence societies in the world, but some use their environment better than others. There is, therefore, the need to unveil the strategies of fostering subsistence agriculture, consequently helping to discover how a nation can efficiently use its agricultural potentials. The notion of the subsistence farmer as one who produces what his or her family needs is largely a fiction. It is in line with this that this paper supports the Wikipedia encyclopaedia that subsistence agriculture is carried out for survival with few or no crops available for sale. In this case, subsistence agriculture should be conceptualised as a farming system. That is, adapting agricultural practices to the conditions of each location and environment.

The place of subsistence agriculture in economic development

The sub-Saharan African economy is growing at the expense of subsistence (Redding). Efforts in the past at developing the economy have resulted in a largely oil based economy and the introduction of the Structural Adjustment Programme (SAP), even though the implementation of this programme failed in virtually all the sub-Saharan African countries where it was introduced. Its aim of liberalising the market was to a great extent achieved (Redding, 1998).

Subsistence agriculture has an important role to play in a country's development, it could provide labour, capital and, more importantly, it would need to supply the food needed for the urban and industrial sector. Typically, women work as members of the household on household controlled land to provide food at subsistence level. The liberalised economic development consequently requires women and youths to work longer hours in cash-crop production resulting in less production of food crops to meet subsistence level and or in pulling young people out of school to meet market production. The additional benefit derived from this is usually pulled into the general economy through undue purchases, which consequently leads to inflation in prices of goods and services, particularly agricultural goods, thus creating a spiral effect on subsistence agricultural production.

Although subsistence agriculture at first may appear to be an impediment for economic growth, it often is the only way for rural people to survive under extremely difficult conditions such as inefficient input, output, credit and labour markets, risks and uncertainties. Under such conditions, subsistence agriculture should not be considered as a passive adaptation. It can ever play an important role in the stabilising of fragile economies (Bruntrup and Heidhues, 2002). Policies need to take these aspects into account and instead of neglecting and even fighting subsistence agriculture, they need to address underlying reasons for drift into subsistence and open viable ways for farmers to increasingly join market economy.

Determinants of subsistence agriculture

According to Rogers (1970), any policy effort that aims at changing subsistence agriculture needs to understand its determining factors. Experience has shown that programmes of direct change designed to reach peasant farmers are likely to fail unless based upon understanding of the values, attitudes and motives of their audience.

In spite of all the compelling forces, the rural people still remain in subsistence agriculture and even induce others outside agriculture to go back to their roots. This, according to Rogers (1970), is because agriculture is ubiquitous and multifaceted, and it also depends on internal and external factors such as:

a. Country internal factors that are given for each country and region such as ecology, climate, history, culture, and international environment (outer circle).

b. The farm-external and country-internal factors such as government policies, institutions, market, etc., which can be influenced by the country itself but are exogenous for the individual household (intermediate circle); and

c. The farm household internal factors, that is factors-endowment and farm family specific characteristics.

Farm household decisions are influenced by all categories of factors and many of these factors are inter-linked and influence each other.

Models of subsistence agriculture

There are various models of subsistence agriculture. These models are considered as basis on which subsistence agriculture is built and provides the instrument for fostering subsistence agriculture. These models are:

1. Subsistence agriculture without risk consideration

a. The propensity for leisure model - The most prominent place among the theories of subsistence production in view of Thorner et al (1966) is the propensity for leisure model by Chayanov. His theory is based on quantitative research on Russian peasant agriculture in the 1920s. The fundamental hypothesis is that peasants have high marginal propensity for leisure even if it means sacrificing additional income, that is, they prefer to reduce the hardship of manual work. This behaviour can lead to a backward-bending supply curve, this basically is why many authors argue that subsistence farmers put up a non-economic behaviour.

b. The hypothesis of satisfying strategy - This emanated from Upton (1982), and it is at an extreme variance from Chayanov's propensity for leisure approach. He said that household work only to achieve a certain minimum level of consumption; they aim at maximising leisure. Such behaviour plays a dominant role in the alternative subsistence agriculture. It automatically leads to a backward bending supply curve. Chayanov's theory according to Durrenberger (1984) is found useful in explaining behaviour in some areas in Africa, particularly the sub-Saharan Africa.

c. Farm/Household model of the new home economies - This is a further development of the Chayanov's model of peasant behaviour. It integrates consumption choices into household's time allocation decision. It maximised the utility of three types of goods: self- and market produced goods, home-produced goods for household consumption such as fuel search, water carrying, cooking, house repair, child raising, etc. and real leisure. In this case Singh et al (1986) say that a backward bending supply curve for agricultural products is possible, if there is no labour markets.

2. Risk and risk coping model

The previous models of explaining subsistence behaviour according to Bruntrup and Heidhue (2002) were basically without explicitly analyzing the role of risk and anti-risk behaviour. The De Janry and Sadoulet (1994) model allows to introduce risk partially by introducing price risk as a shift of the price bands. But the issue of risk is said to go beyond price risks.

Strategies to cope with risks infiltrate the whole life of rural poor; risk is said to be a constituent element of subsistence economies. A common and consistent result of researches on risk behaviour of farm/household have shown that subsistence farmers are extremely risk aversive (Binswanger 1980; Antle, 1987). This can be generalised for most human behaviour in situations where the consequences of risk are serious or life threatening.

Fostering subsistence agriculture

Various strategies are required to foster subsistence agriculture in sub-Saharan Africa taking into consideration the nature of agriculture and its audience in this region.

A great percentage of the rural population in sub-Saharan Africa are involved in agriculture as a primary occupation. However, they still cannot meet their food requirements as a result of the demanding economic needs. This has compelled them to commercialise their agricultural production efforts at the expense of self- subsistence, in order to enhance subsistence agriculture even to a sustainable level, the following strategies are necessary:

a. Multiple occupation

Farmers engaged in agriculture are better-off when involved in other income generating activities to cater for the economic needs of the household, while the farm produce is reserved for subsistence consumption and only left overs from household consumption are sold out. Generating income from other activities reduces the pressure on farm produce, which presently cannot even sustain the household. Such pressure has resulted in premature harvest, consumption and/or sales of farm produce which consequently reduces the nutritional and market value of the produce.

In a situation where the farmer is encouraged to produce enough for subsistence consumption, because there are other sources of income, subsistence agriculture is not only enhanced but result oriented (Wikipedia encyclopaedia).

The Peruvian economist Hermando de Soto has argued that one obstacle to industrial development is that subsistence farmers cannot convert their work into capital which can be used to start new businesses and trigger industrialisations. He argues that these obstacles exist often because subsistence farmers do not have clear titles to land, which they work, and to the crops, which they produce.

b. Home gardening

This is a long standing subsistence agriculture production technique that should be revived and encouraged. The technique was phasing out as a result of urbanisation, land problems, poor attitudes, etc. Reactivating a positive attitude toward backyard farming and home gardening to a great extent will enhance subsistence agriculture. Home gardening involves land saving intensive cultivation techniques that enhances soil fertility over a long period of time through the use of domestic waste and dungs, from semi-intensive livestock and poultry management systems.

c. Gender consideration

In sub-Saharan Africa, agricultural production is gendered to the extent that when men farm, they often farm principally for the market, whereas when women farm, they farm principally for household consumption. This shows the significant role women play in subsistence agriculture. The role of many women is that of partnership with the husband, though not on equal basis, in providing for the home while the men work for money to meet other needs. In many areas of sub-Saharan Africa, most of the tasks, if not all, in subsistence food production were left to the women. Such tasks varied according to the size of the farm holding. Thompson (1986) discovered the leading role of women in subsistence agriculture by stating that even poultry and the products often come immediately under the care and management of women. These the women use in balancing the nutritional value of the domestic meals and sell some of the left over to earn though meagre incomes but which could meet trivial necessities in the home, such include salt, sugar, spices, oatmeal, etc. Taking advantage of this gender differentials in subsistence agriculture will definitely translate to gender comparative advantage in enhancing subsistence agriculture.

The role of women according to Buckland (1996) is often socially defined and shaped by the traditions and beliefs of a particular culture. All African cultures support the involvement of women in agriculture as primary producers for the family and the domestication of the home. Women are also the major labour forces in processing and storage of agricultural products. Agricultural extension services should be directed at educating the women on the need to prioritise home consumption even at the face of the need to commercialise, as a poorly fed family has very limited capacity to foster commercial agriculture. Subsistence agriculture in this case becomes a prerequisite to enhancing other efforts in meeting other economic needs.

d. Youths involvement in subsistence agriculture

Youths being the energetic age bracket should be encouraged in producing food for themselves and their home by engaging in subsistence agricultural production. When production techniques are made as a way of life through emphasis on passing down indigenous knowledge systems in combination with the provision of inputs such as fertilisers, improved seeds, cuttings, chemicals animal feeds, etc. that make subsistence agriculture exciting, subsistence agriculture is made easy. The reactivation of school to land programmes for the youths is also necessary.

In a study carried out by FAO (1997) on youths involvement in post-harvest activities in rural communities of Bolivia, Philippines and Vietnam, it was found out in the three countries that older youths involved in commercial agriculture were highly vulnerable to discontinuing their education once they start receiving a little salary, while the younger youths have low school grades. Generally, youths' involvement in commercial agriculture was found to have negative effects on their education, unlike the subsistence agriculture which gave enough time for school and leisure and also had the transfer of indigenous knowledge as its component. This same results can also be said of African countries particularly the sub-Saharan region.

e. Use of Extension Services

Cottrel - Dormer (1967) observed that in the early stages of development, extension services were usually organised by Kinship groups. In addition to cultivate crops and animals, the family also supplies its own seeds and tools, and most importantly, they put aside subsistence requirements in the grain store. Subsistence at this time was given prime attention. Extension can include subsistence agriculture as an approach adopted to reach out to farmers while attempts are made at resuscitating the kinship family extension system.

f. Government legislation and policies

Through government legislation and policies, the subsistence agriculture can be made as a common way of life of the people. Policies that will encourage subsistence agriculture through the provision of land and facilities for homes, school, organisation, etc., production to meet their agricultural needs are very necessary. The economic policies and decisions are usually those, which encourage large scale production at the expense of subsistence and consequently impacting negatively on the instruments of subsistence agriculture.

Problems of subsistence agriculture

The subsistence farmer is faced with the problem of deciding a strategy in using the land. Most farmers will measure their productivity in meeting simple cash needs against numbers of hours work and food requirements. Very often, women grow food crops and men cash crops, and men value women's time as less valuable than their own. All the same, as commodity prices decline, the rewards from cash crops on small areas of land are so small that most subsistence farmers consider it non-economic to grow them. In this case, men need to be more educated through proper extension service on the need to prioritise subsistence agriculture as a pre-requisite to market agriculture. Thus if people take the least effort option to survive, they create roles and constraints to make life more complicated and difficult.

Other problems of subsistence agriculture as identified by Bruntrup and Heidhue (2002) are:

1. Poor research of subsistence agriculture due to its perceived low productivity and lack of dynamism to act as motor of economic development.
2. Subsistence farmers' behaviour often appears mysterious to economists; they seem to behave in ways not consistent with the principle of economic theory or irrationally, not reacting to price changes or incentives.

3. The science of economics and its analytical tools are mostly based on existence of markets, which are outside of the range of subsistence.
4. Statistical data about subsistence are usually not available or unreliable when available.
5. Subsistence farming seen as, or derided as, traditional and resistant to change and innovation, thus not a preferred target grip for development practitioners as policy makers.

Summary and conclusion

Subsistence agriculture as seen in this paper goes beyond just producing for household consumption. However, household consumption is the priority of subsistence farming, consequently, such agricultural technique must be able to provide adequate food in quantity and quality to meet the domestic and health requirements of the rural populace by ensuring proper nutritional value. It is after meeting this level, that subsistence agriculture is said to create the basis for commercial agriculture. This requires the strategy of involving and repositioning the stakeholders in subsistence agriculture in such a way as to foster subsistence farming. Major stakeholders include the subsistence farmers, mainly women, youths and men, government and extension services.

The male farmers should appreciate the subsistence role of the women farmers. Policies-orientations directed against subsistence agriculture will be counter-productive in rural sub-Saharan Afric unless commercial agriculture is built within the framework of a successful and sustainable subsistence agriculture. Consequently, the potential benefits of agriculture are maximised; there is the promotion of technological change in subsistence food production along with commercial crop production for household food security in areas with risky food markets.

Finally, subsistence agriculture can be made a component of rural development strategy which involves improving the living standards of the mass of the low income population residing in rural areas and further making the process of their development self sustaining.

References

Antle, J. M. 1987. “Economic Estimation of Producers Risk Attitudes” *American Journal of Agricultural Economics* 509 –522.

Binswanger, A. P. 1980. “Attitudes toward Risk: Experimental Measurement in Rural India.”*American Journal of Agricultural Economics* 395– 407.

Bruntrup, M. and F. Heidhues. 2002. "Subsistence Agriculture a Development: its Role in Processing of Structural Change." *Institute of Agricultural Economics and Social Sciences in the Tropics and Sub-tropics,* ed. *Forschung zur Entwicklungsökonomie und -politik - Research in Development Economies and Policy.* Discussion Paper 1, 2002.

Bucklard, L. and J. Haleegoah. 1996. *Gender Analysis in Agricultural Production.* Ibadan: IITA, Research Guide 58.

De Janvry, A. and E. Sadoulet. 1994. "Structural Adjustment Under Transaction Cost:" Heidhues and Knerr (eds). *Food and Agricultural Policies Under Structural Adjustment.* Frankfurt. pp 137 - 165.

Durrenberger, E. P. (ed.). 1984. *Chayanov, Peasants and Economic Anthropology.* New York.

International Institute of Rural Reconstruction. 1998. "The Role of Children in Post-harvest Handling in Asia." A Report in FAO, January 1998. IIRR Philippines.

Rogers, E. M. 1970. "Motivations, Values and Attitudes of Subsistence Farmers toward a Sub-culture of Peasantry." Wharton (ed.). *Subsistence Agriculture and Economic Development.* London. pp 111 - 135.

Redding, Sean. 1998. *Structural Adjustment Arid the Decline of Subsistence Agriculture in Africa.*

Singh, I., L. Squire, and J. Strauss. 1986. *Agricultural Household Models: Extensions, Applications and Policy.* Baltimore.

Thompson, Marilyn. 1986. *Women of El-Salvador. The Price of Freedom.* London. Zed Book Ltd.

Thorner, D., B. Kerblay, and R. E. F. Smith. 1966. "A. V. Chanayov." *The Theory of Peasant Economy.* Homewood/Illinois.

Upton, M. 1982. "Grundlagen der bäuerlichen (und grobbetrieblichen) Betriebsorganisation und der Betriebsführung in Entwicklungsländern.", V. Blackenburg (ed.). *Sozialökonomie der ländlichen Entwicklung, Handbuch der Landwirtschaft in den Enwicklungsländern* 1, 101 – 124. Stuttgart.

von Bruan, Joachim and Gileen Kennedy (eds.). 1994. *Agricultural Commercialization, Economic Development, and Nutrition.* The John Hopkins University Press, 2715. North Charles Street, Baltimore, Md., USA.

FARMING SYSTEM APPROACH

Ngozi Edith Anozie

Abstract

The need for a farming systems approach which involves identifying farmers inputs, outputs, development of interventions to resolve constraints, as well as encourage the adoption of farming alternatives that have been proven to be technically, socially and economically feasible and compatible and acceptable to the farmers must be encouraged. Also a multi-disciplinary research team effort consisting of sociologists, economists and biological scientists are needed to ensure the realisation of the full benefits of the research process by the farmers.

There is also need to recognise the farmers' cultural and agrarian ecological conditions to enable the tapping of the farmers indigenous knowledge which will help to integrate farmers into research initiation, planning and implementation. Government should initiate policies that will enhance rural agricultural development through the establishment of social amenities such as rural feeder roads, electricity, water supply, communication and markets in the rural areas. There is also need to ensure good governance that will foster peace and economic empowerment for the rural poor farmers. Women should be given access to political, economic and social incentives in view of their prominent role in rural agriculture. Women's access to land ownership and other support services must be legalised by the government to boost their productivity in the agricultural sector.

Introduction

The domestication of plant and animal species started at about 12000 years ago and this marked a momentous transformation in human subsistence strategies. Agricultural societies take many forms influenced not only by habitat and ecology but also by cultural and political factors. Subsistence agriculture is a form of agriculture which relies mainly on localised inputs and is oriented primarily towards household provisioning rather than the production of surpluses or investment. The agricultural sector dominates the economies of most sub-Saharan African countries. The sector accounts for about 42% of Gross Domestic Product in low income countries.

In spite of the dominant role that agriculture plays in the economies of most developing countries, food production has been on the decline due to a number of factors. The increased population density have impacted pressure on cultivable land leading to a drastic reduction in the fallow periods - which is the traditional means of nutrient conservation of degraded soils by the subsistence farmer. Besides affecting aggregate food supply, soil degradation also diminishes agricultural income and economic growth.

Historical as well as socio-economic evidence suggests that farmers often respond actively to environmental degradation by modifying their farming systems or practices through land improvement strategies. Conservation farming is being practised in some developing countries. Diversification into high value perennial crops is protecting soils in Africa (Sara and Yadau, 2001). An effective response to land degradation calls for improving the incentives for farmers to care for their land and improving their access to the knowledge and inputs required for proper care. There is need for a policy action to increase research and technology development for land management as well as improve the spread of information on user-friendly information systems for farmers. There is also need to promote land improving investments through technical assistance and new financing arrangements suitable for subsistence farmers. There is also need to encourage long term land improvements by securing property rights and rights of access to natural resources especially to women farmers and to develop long term planning systems for sustainable land use that involve key resource user groups. Government should also improve the economic environment per farmers in the rural areas by developing market infrastructure, correcting distorted price incentives and encouraging rural income growth and diversification.

Since most of the developing countries are at a very early stage of agricultural technology, there is need for increased support for agricultural research to raise productivity beyond the subsistence level. In the labour constrained systems such as in West Africa and Central Africa, technologies that permit rapid increases in labour productivity are essential (Delgado and Pinstive, 1993). Also yield enhancing technologies are probably urgent for the land constrained systems which tend to predominate in Africa, since the scope for continued expansion of production by bringing new land into cultivation appears to be limited.

Biotechnology also offers a significant potential for increasing the returns to agricultural research. There is urgent need to increase support for agricultural research to emphasise the development of technologies for sustainable farming and natural resources management and to forge links with centres of biotechnology research. Government must play an active role in the provision of extension services capable of disseminating relevant information to farmers and encouraging non-governmental organisations (NGOs) working on research and extension problems.

There is also need for women to be fully accorded their full rights to land and other economic resources needed to boost farming. Demonstrating the economic importance of agriculture, particularly in development countries, is not difficult. The combined approach of developing pools of local expertise and sharing experiences widely has accomplished three important tasks in the development of a more sustainable agriculture:

1. It has generated and adapted locally-based alternatives that are easily incorporated into the ecology of each particular area, increasing agro-ecological diversity and resilience.

2. It has spread simple, adaptable technologies at low cost to thousands of farmers and has improved the capacity to innovate.

3. It has developed farmers' social and agro-ecological capabilities (ILEA Newsletter 13.3, pp 12-13). Professor R. Evenson and Mr. Jaycox stressed that Africa remained a top priority for the World Bank in spite of new demands and commitments in other parts for the world. This priority is reflected in the consistently high level of World Bank assistance to Africa and the presence of resident missions in many of these countries. Within this overall priority for Africa, agriculture is accorded priority for social and economic reasons, such as the percentage of Gross Domestic Product (GDP) from agriculture; its contribution to exports; employment; the role of agriculture and farmers in natural resource management and so on.

Strategies for agricultural development

Agriculture remains the principal occupation of the majority of people in Africa. A successful strategy must be based on the assumption that agriculture is a positive factor in long-term sustainable agriculture and that research will enhance the contribution of crop/livestock to sustainable agricultural development (World Bank, 1990). Strategies should thus embody the following concepts: adequate economic returns to crop/livestock farmers; maintenance of natural resources and productivity; minimal adverse environmental effects; optimal production with minimal external inputs; and satisfaction of human food and income needs as well as of rural families social needs. Given the complexity of agricultural management, a farm system approach to developing and implementing research strategies is essential. To ensure that research undertaken addresses farmers' perceived needs, such an approach must: characterise the farming system - identifying inputs, outputs, potentials, design alternative systems; evaluate interventions and alternative systems, develop interventions to resolve constraints, and elaborate alternatives that have proven to be technically practical, economically feasible and socially acceptable (Cleaver, 1993).

The research process should begin with the careful identification of problems. Systems description and constraint analysis are particularly valuable for informing scientists about the real problems faced by farmers. An interdisciplinary team effort, combining efforts of socio-economic and biological scientists, is required to ensure that the full benefits are realised from research to the development stage. In the planning and priority-setting stage, the social scientists can provide exact projections of costs and benefits while helping to address equity issues that may arise, such as gender and reconciling the interests of the poorer farmers with those of the more prosperous producers.

As research progress, the monitoring of socio-economic factors improves the probability of favourable social and economic impacts. The research strategy recommended for sub-Saharan Africa identified the priorities of genetic improvement, livestock management, crop/livestock farming systems, natural resources and policy (Winrock International, 1992). Priorities for research were based on agro-ecological, biophysical and socio-economic constraints.

The need for expanded agricultural production to provide food for today's population is increasingly affected by the fear that the eroding natural resource base will not meet the food needs of future generations. Farmers are already cultivating the better grazing lands, thereby limiting pastoral herbs to marginal lands even those marginal lands are increasingly threatened by the expansion of cultivation. If agricultural development is to be sustainable, research is needed to intensify and increase productivity from the better endowed and more robust lands and to improve the management of soil, water and vegetative resources on the more fragile lands (Winrock International, 1992).

World Bank's report entitled “Sub-Saharan Africa: from Crisis to Sustainable Growth, *Long-Term Perspective Study*” (LTPS, 1989) concluded, that Africa's economies needed to grow by at least 4 to 5 percentage or to meet the employment and income needs of their growing populations. Since the primary source of this growth can only be agricultural production, the agricultural growth rate in these countries would also have to be 4 percent per annum. Without such agricultural growth rate, the industrial sector will not be supplied with adequate raw materials to reach its targeted 5 to 7 percent rate of growth. Agriculture is the major source of raw material for industry, as well as a significant purchaser of the countries' manufactures and services, since farmers constitute the majority of the population and of the employed. The stagnation in agriculture has led to a worsening incidence of poverty and food availability in the rural areas. The current production systems are unable, on the one hand, to absorb the increment labour, and on the other, to prevent the progressive destruction of the underlying resource base.

The crucial involvement in spreading improved technologies to farmers, and in advising them on how to better manage their natural commercial crop and livestock enterprises can, in due course, be provided by private sector agri-business, public sector extension will continue to be important for a long time for food crops and for poor farmers. Public sector extension must increase its efforts towards serving women farmers, addressing environmental issues, in providing a choice of technologies and in establishing formal links between research and farmers groups.

Agrarian-cultural and agrarian-ecological strategies

Farmers' specific entry prompted out that technologies are very important as they bring fast economic results, make further investments possible and create the enthusiasm, motivation and self-confidence necessary to further develop more integrated agro-ecological agriculture (LEISA) and sustainable livelihoods. Zero tillage farmers in Brazil, for example, have realised that herbicides based on no-tillage is far from sustainable and are now developing a broader approach, namely conservation agriculture. In all these cases, farmers have learned to use local natural resources and ecological processes in more efficient ways, to reduce the use of costly external inputs and to intensify agriculture ecologically.

Thus the cultural systems gradually evolve into integrated systems. In general, ecological intensification also leads to higher ecological sustainability and resilience. In most cases (except in organic agriculture), farmers do not exclude agrochemical means completely, but use them cautiously in a way that is appropriate for their specific economic conditions. Reduction of the use of external inputs and development of integrated agriculture, however, may only happen after farmers have experienced the negative impacts of mono-cropping.

The drive towards agro-ecological agriculture is strongest in subsistence conditions where agrochemical are too expensive to use, credit is not available and farming has to satisfy a wide array of subsistence needs in commercial agriculture, farmers and researchers are trying to develop integrated pest, disease, weed, soil, and water management. The development of agro-ecological agriculture and local healthcare, off-farm income generation and strengthening of local value and knowledge based decision-making can be mutually supportive and serve as crucial elements for successful development of sustainable livelihoods.

The adverse effects of tropical deforestation, such as climate change, loss of bio-diversity and soil degradation have been well publicised. However, agriculture production systems based on humid savannahs can be quite productive, proving control of major diseases, especially trypanosomiasis, is effected. Increased population pressure will almost certainly expand exploitation of the humid zone.

The fallow period of traditional slash-and-burn farming systems is already being shortened in many regions, leading to the pernicious loss of soil resources. Research on nutrient-cycling involving the use of animal manure, usually in combination with chemical fertiliser, may offer an alternative to slash-and-burn practices in the humid zone.

Political development aspect

Farmer support for resource conservation has been shown to depend greatly on government policies. The effects of credit, pricing, and monetary policies on resource use and the environment, especially for the mixed production systems, are therefore important points to consider. Policy-induced distortions in financial markets, for example, result in small farmers' having to pay high real interest rates. Many farmers who are unable to borrow or meet the repayments on borrowed funds seek refuge in common access areas, which are often susceptible to environmental degradation. Policy-induced distortions in agricultural commodity markets also result in farmers receiving low prices for the agricultural products, thereby discouraging them from investing in natural resources management.

However, most agricultural policies still focus solely on increasing production, usually associated with packages of external inputs and technical interventions. In many cases, these policy farm works are the principal barriers to the spread of more integrated, sustainable and productive agricultural systems (FAO, 2001).

Political aspect

A community-base programme or a grassroots organisation expands the number and the type of its activities e.g. from agricultural production to health, nutrition, credit, training, literacy, etc., while the organisation moves beyond service delivery towards empowerment and change in structural causes of under-development. This usually involves active political involvement and the development of relations with the state.

The demonstration to the national government shows the importance of co-operation in a region to raise more funds for the decentralisation process makers' way for defining and implementing common policies that promotes sustainable agriculture in a local/regional content, e.g. tax reduction for projects, laws to punish pesticide used, etc. And to facilitate designing and implementing common programmes for sustainable agriculture in watershed and mobilising institutional resources (human, financial, physical) and to multiply its resources through social and private resources.

Economic aspect

Social capital among social, public and private institutions can increase the competitiveness of a region in the country and makes it possible for new loans from banks to farmers' organisations (social guarantees among farmers can serve as replacement/alternative to collateral requirements in the absence/lack of property rights to land) thereby making it possible to design and implement new strategies to reduce the risk of markets (new role among municipal governments, farmers organisations, NGOs, local banks and private sector can increase the efficiency of institutional and create co-operative guarantees to attract investments and loans.

Social aspect

Modern management emphasises the need for indicators to measure progress. We have found that seeking universal indicators should be chosen to fit the local time and place. The communities need to agree on the direction they want to go and a shared vision helps internal collaboration. This can evolve from immediate concerns or from underlying causal influences over time as observed in annual reviews.

The only way a whole community can improve overall welfare is when the groups which have the greatest concentration of priority problems are helped to get better living conditions and lifestyle. And the depending produces victimisation and vulnerability to be controlled by outsiders. In many instances, outsiders assistance has been great in satisfying the egos of givers but has been terrible for the self-reliance that happens when capacity building is a primary goal. Development cannot be bought. And in order to accept, only if they do not remove control, for the criterion that assistance must give priority to promoting equity.

The role of women in agriculture

Women are responsible for at least 70 percent of food staple production in Africa (World Bank, 1989). They are also important in other agricultural activities, including food processing and marketing cash cropping and animal husbandry. Women's involvement is significant not only in terms of their labour input, but also in terms of labour decision-making authority.

In fact, as more men migrate to cities and other countries for work, increasing numbers of women are becoming heads of households, managing farms on a day-today basis. To support their extensive and multifaceted roles in agriculture and to enable them to respond to market incentives more efficiently, women need effective agricultural extension services.

The evidence shows, however, that despite a growing awareness of the need to reach women farmers, agricultural extension services are generally geared toward male farmers. This is sometimes by design, but more often, by default. Bias toward male farmers is evident in the delivery of extension, which is generally provided by male extension agents to men, on the assumption that the extension message will "trickle across" to women.

Unfortunately, the evidence clearly shows that it often doesn't. It is also evident in the activities of male farmers rather than the much wider range of agricultural activities of women. Such an approach ignores the unique workload, responsibilities and constrains of women farmers and results in a highly inefficient use of resources, not to mention sub-optimal levels of agricultural producing.

All small-scale farmers, regardless of gender, face constraints, but the focus here is on women farmers in order to foster a better understanding of the particular gender-related barriers confronting women and the strategies needed to overcome them. And the attention is concentrated on sub-Saharan Africa in view of the crucial role of women in agriculture throughout the subcontinent. The recommendations have been gleaned from the experiences of African governments, the World Bank and other donors, and researchers. The ongoing pilot programs have provided useful guidance about what can work to integrate women fully into the agricultural extension system and what problems are likely to emerge in different socio-economic environments (Office of Technology Assessment, 1988).

Women's role in agriculture

1) Women provide most of the labour and make the key decisions for many agricultural activities. This applies not only to food production, long recognised as primarily a women's activity, but also increasingly to other agricultural activities ranging from processing and marketing to cash cropping and animal husbandry (Buvinic and Lycette, 1988).

The data on the role of women in agriculture in Africa is fragmented, but the international labour organisation estimates that over 95% of the rural women are small-scale farmers, who produce most of the food and bear the burden of day-to-day family subsistence. According to the diagnostic surveys of the agricultural extension zones, women perform almost all the cultural operations in food production. Such operations as bush clearing, burning, ridge/mound making, planting, fertilisation, weeding, harvesting, storage, processing and marketing are carried out by women. Women also have sole responsibility for cultivating compound farms (or gardens), where continuous cropping is done beside the tasks of the household. It was against this background of active participation in food in Imo State that ADP decided to give full and adequate technical and financial support.

2) Official underestimates: Standard data sources have been criticised for underestimating female workers especially those who are unpaid, and for undervaluing the worth of female produced goods and services. The fact that women's involvement in agriculture is typically much greater than official figures suggests has enormous implications for the type and extent of the agricultural support services they require (Jiggins, 1986).

3) Female heads of households: A growing phenomenon across Africa is the female-headed household. Definitions of what constitutes a female-headed household vary, but evidence shows clearly that the number of households managed by women on day-to-day basis is growing rapidly -- largely because men are migrating to large farms, cities or mines for work. For example, female-headed house-headed households are particularly common in the Southern Africa Development Co-ordinating Conference (SADCC). It is estimated that over half of rural households are headed by women. Available data show that in rural Africa households headed by women are poorer than those headed by men, because they tend to have fewer resident workers, more dependants and smaller land holdings (Tinker, 1981).

4) Improving women's productivity: The debate has evolved from the **extent** of women's participation in agricultural activities to a search for practical solutions as to **how** to develop agricultural support services that can effectively

make women farmers more productive. Given the evidence of women's significant responsibility for and contribution to agricultural production in Africa, one would expect women to be participating fully in developing the extension messages and in delivery those messages to farmers. Yet the evidence show this not to be the case. Despite a growing awareness of the need to reach women farmers, extension services are still skewed towards males (Berger et al, 1984).

Conclusion

Strategies aimed at improving farmers productivity must be sustainable, environmentally friendly, adaptable and culturally acceptable by the target groups. The research process must involve farmers at the initiation, and developmental stages to ensure smooth adoption of new technologies by the farmers. Such new technologies must be integrated in the already existing farmers' indigenous knowledge.

There is need to develop the rural infrastructure to foster agricultural productivity. Also there is need to ensure peaceful co-existence in all communities. Good governance and establishment of good marketing facilities will improve farmers income and boost their productive potential.

Finally, women should be socially, culturally, politically and economically empowered as a major stakeholder in the farming industry.

References

Buvinic, M., and M. Lycette. 1988. *Women, Poverty and Development in the Third World.*

Berger M., and V. Delancey. 1984. "Bridging the Gender Gap in Agricultural Extension," *International Centre for Research on Women*, Washington, DC.

Cleaver, Kevin. 1993. *A Strategy to Develop Agriculture in sub-Sahara Africa and a Focus for the World Bank.*

Delgado, C. L., and Anderson Pinstrup. 1993. *Agricultural Productivity in the Third World Patterns and Strategic Issue. A Keynote Address to the 1993 America Agric-Economic Association.*

FAO, 2001. *Conservation Agriculture - Case Studies in Latin America and Africa.*

Ileaia Newsletter, Oct., 2001. Vol. 17, No 3. *Magazine on Low External Input and Sustainable Agriculture.*

Jiggins J., 1986. *Gender Related Impacts and the World of the International Agricultural Research Centre.* Washington, DC.

OTA, 1988. Office of Technology Assessment Enhancing Agriculture in Africa. Congress of the United Stated, Washington, DC.

Ohuegbe C., 1989. "Women in Agriculture Programme in Imo State." Paper Presented to World Bank Workshop in Agriculture. Ibadan, Nigeria.

Orlandao Florida. July, 1993. International Food Policy Research Institute Pre-conference Workshop. Post Green Revolution, Agric Dept.

Scherr, Sara J., and Yadav Satye, 2001. *Land Degradation in the Developing World, Issues and Policy Options for 2020 in the Unfinished Agenda.* IFFRI. Washington, DC.

Tinker, I., 1981. *Energy for Essential Household Activities.* Washington, DC.

World Bank, 1989. *Sub-Sahara Africa from Crisis to Sustainable Growth, a Long-term Perspective Study.* Washington, DC.

—— 1989. *Kenya. The Role of Women in Economic Development.* World Bank Country Study, Washington DC.

—— 1990. *World Development Report 1990, Poverty.* Oxford, UK. Oxford University Press.

World International, 1992. *Assessment of Crop/Livestock Agriculture Products in Sub-Saharan Africa.* Arkansas, Morilton.

STRATEGIES FOR FOSTERING SUBSISTENCE AGRICULTURE

Cyril Ifeanyi Duruigbo

Abstract

The research strategy shall involve a farming systems approach, which seeks to identify the farmers agroecological, environmental, social, economic, cultural and political problems. The need for collaborative research between the farmers and scientists is essential just as a holistic and interdisciplinary team effort is crucial in tackling the diverse nature of the farmers problems.

The agroecological approach shall mobilise farmers' indigenous knowledge system and properly tap the development of new technologies. Farmers should be at the centre of the technology design process, formulation and its implementation. Sustainable agricultural practices must be integrated into the agroecological processes to optimise both human and natural capital.

Good governance must ensure the availability of social infrastructure as well as peaceful co-existence to enhance food production.

Governments should develop social policies that will ensure good education, healthcare, good road network, efficient marketing and storage systems as well as good communication systems for the rural poor subsistence farmers in order to boost food productivity.

Women should be adequately and economically empowered by giving them access to physical and human resources just as men.

There is need to dismantle all economic, cultural, political and social practices that prevent women from reaching their full potential.

Introduction

Subsistence agriculture is farming carried out for survival, with few or no crops and livestock available for sale and is usually organic due to lack of money to buy industrial inputs such as inorganic fertiliser, pesticides and genetically modified seeds. It is, however, not always sustainable agriculture due to the slash-and-burn system practised as well as the use of crude tools for farming.

Historically, most farmers were engaged in these practices for subsistence and this is still the case in the poorest developing nations where over one billion people live under one US dollar per day and two billion live on under five dollars per day. (World Bank 2000, Pinstrup-Andersen et al, 1997). Globally about 90 percent of developing world's poor people live in Asia or sub-Saharan Africa. Of the total number of poor people living below one dollar per day, 75% or 0.9 billion live and work in rural area where subsistence agriculture is a major means of livelihood (IFAD ,2001).

A significant percentage of the rural poor live in less-favoured areas that are challenged by difficult agro-climatic conditions such as poor soil, low and unstable rainfall, steep slopes, inadequate infrastructure and support services such as good roads, irrigation, markets, research and extension, credit facilities, schools and health centres. The rural poor are predominantly smallholder farmers and landless agricultural workers. In the developing countries, about half the rural population lived in smallholder farm household in 1988 and one-quarter lived in landless labour households (Jazairy et al, 1992).

Smallholder or subsistence farmers are getting smaller and more numerous in most parts of the developing world (Hazell et al, 2000). Subsistence farmers are also diversifying their livelihood strategies and increasing their share of non-farm income (Reardon et al, 1998; Carney, 1998). Such diversification could reflect worsening impoverishment and desperation as land becomes increasingly scarce.

Diversification is more likely to be associated with greater impoverishment when increasing land scarcity occurs in conjunction with slow agricultural growth, stagnant national and regional economies and falling wages (Hazell and Reardon, 1998), which characterises most developing countries in South Asia and sub-Saharan Africa.

In the past, traditional agriculture had been compatible with the level of population and the ecological environment, when long bush fallows were effective in restoring soil fertility for the level of crop yields and intensity of cropping. The increased population density has mounted pressure on land and thus resulted in drastic reduction of the fallow periods in many cases; in some countries fallowing has completely disappeared and has been replaced by intensive land cultivation - albeit with low use of inputs due to the farmers' inability to purchase improved inputs, reduced crops yields and degraded soils is the consequence. The effect of population pressure has also reached marginal lands which fall easy prey to rapid soil degradation, since the farmers have not acquired the necessary soil management expertise to cope with these challenges. Degradation of natural resources is rampant in man-made resource-poor areas of developing countries particularly those areas with fragile soils, irregular rainfall, high population concentration and stagnant productivity in agriculture. Natural resource degradation is also occurring in agricultural areas exposed to misuse of modern farming inputs. While natu ral resource degradation is often a consequence of poverty, it also contributes to poverty in many subsistence farming environments. In order to transform the subsistence agriculture in the developing countries considering the socio-cultural, political, agroecological and environmental constraints affecting productivity there is need to adopt a broad-based strategic initiative that will not only revitalise food production on a sustainable basis but will ensure environmental safety and preservation of the natural resource base. These strategies will include the following:

Research strategies

Eco-systems research suggests that high productivity and stability of ecological systems are associated with high degrees of cohesiveness (Hartman, 1973), functional diversity and complexity (Egger 1979). Eco-farming involves finding the optimum rather than the maximum cohesiveness and functional diversity for each specific location (Kotschi, 1981). The research process should begin with an intimate knowledge of local conditions of the subsistence farmer. This is because farming in Africa, Asia and South America is practised under a wide range of natural, economic, cultural and political conditions. In view of the site-specific nature of many agricultural problems, widely applicable solutions are rare. Technologies and practices must be adapted to meet the needs of specific location. Site specific solutions can be found only if development planners, policy makers, researchers and extension workers strive to collaborate with farmers and assist them in their efforts to develop the most appropriate technologies and practices for their particular conditions, rather than trying to implement pre-conceived ideas and methods that may have been successful in another context but are ill-adapted to local circumstances. There is also need for an interdisciplinary understanding of agro-ecosystems. The environment is not confined to natural conditions alone, it encompasses the entire complex of physical, economic, social and cultural conditions including their relations and interactions. There is need for research on social and economic aspects of the subsistence farmer conducted simultaneously with biological or agricultural science in all attempts to understand and influence the evolution of eco-systems.

An interdisciplinary team effort, combining efforts of socio-economic and biological scientists is required to ensure that the full benefits are realised from research to the development. In the planning and priority-setting stage, the social scientists can provide ex ante projections of costs and benefits while helping to address equity issues that may arise such as gender and reconciling the interests of the poorer farmers with those of the more prosperous producers. While some types of commodity improvement work seem vital for less favoured areas, improving drought tolerance, yield response to scarce plant nutrients, food nutrient content, pest and disease resistance, livestock health and productivity improvement will first deal with improved natural resource management practices and technologies in areas of poor soils and harsh climate.

The poor infrastructure and market subsistence farming communities make the use of high levels of external inputs uneconomic placing a premium on low external input technologies. Low external technologies are typically labour intensive both seasonally and in total, thus constraining their use. Fallows and green manures tend to keep land out of crop production, while composting and manuring compete with household energy use for scarce organic matter. The challenge is to develop low external input technologies that boost labour and land productivity.

Agroecological approach and mobilisation of farmers' indigenous knowledge

Another strategy needed for fostering subsistence agriculture is to acknowledge the farmers indigenous knowledge system. In many farming systems practical ecological knowledge has been accumulated by generations of rural farmers who often undertook deliberate efforts to improve or protect the value of life-supporting resource and to ensure reasonably secure long-term viability (Beyer, 1980). Millions of small-scale subsistence farmers throughout the tropics are professionals and thus should be recognised and treated as such (Chambers, 1980). Scientists, policy makers and agricultural advisers must learn from local farmers because their expertise represents the single largest knowledge resource not yet mobilised and tapped in the development enterprise (Hatch, 1976). The learning process involves recognising and understanding indigenous eco-farming systems and technologies so as to promote, develop and extend this knowledge to other farmers operating under similar conditions. It also involves placing farmers and their families at the centre of the technology design process and obtaining their contribution or input to problem diagnosis as well as technology testing and evaluation. Recently the on-farm adaptive research processes have accommodated farmers in the participatory technology development. A number of farmers, non-governmental organisations (NGOs) and analysts propose that instead of the capital and input intensive approach which relies on ever-increasing fertiliser and pesticide use to boost food supplies, developing countries should favour an agroecological model which emphasises biodiversity, recycling of nutrients, synergy among crops, animals, soils and other biological components, regeneration and conservation of resources.

Agro-ecology which involves the use of ecological principles for the design and management of sustainable and resource conserving agricultural systems offers several advantages over the conventional agronomic or agro industrial approach. Firstly, agro-ecology relies on indigenous farming knowledge and selected modern technologies to manage diversity, incorporate biological principles and resources into farming systems to increase agricultural productivity beyond the subsistence level. Secondly, agro-ecology offers the only practical way to restore agricultural lands that have been degraded by conventional agronomic practices. Thirdly, agro-ecology also provides for an environmentally sound and affordable way for small-scale subsistence farmers to intensify production in marginal areas.

Moreover, it has the potential to reverse the anti-peasant bias of strategies that emphasise purchased inputs as opposed to the assets that subsistence farmers already posses such as their low opportunity costs of labour.

Data show that over time agroecological systems exhibit more stable levels of total production per unit area than high-input systems, produce economically favourable rates of return, provide a return to labour and other inputs sufficient for a livelihood acceptable to subsistence farmers and their families; ensure soil protection and conservation and enhance agro biodiversity. The question now is despite the increasing evidence and awareness of the advantages of agro-ecology: why hasn't it spread more rapidly and how can it be multiplied and adopted more widely? Clearly technological or ecological intentions are not enough. Major changes must be made in policies, institutions, research and development to make sure that agroecological alternatives are adopted, made equitably and broadly accessible and multiplied so that their full benefit for sustainable subsistence food security can be realised. Existing subsidies and policy incentives for conventional chemical approaches must be dismantled, and institutional structures, partnerships and educational processes must change to enable the agroecological model to blossom. In addition, participatory, farmer friendly methods of technology development must be incorporated. The challenge is to increase investment and research in agro-ecology and scale up projects that have already proven successful, thereby generating a meaningful impact on the income, food security and environmental well-being of the world's increasing population, especially millions of poor resource farmers that are yet to embrace modern agricultural technology.

Furthermore, sustainable agricultural practices which make the best use of nature's goods and services, integrating agroecological processes into food production as well as minimising negative side effects on environment and health offer a broad based strategic approach in fostering subsistence agriculture. This is because it makes the best use of natural capital as well as human capital by building on the skills and knowledge and ingenuity of farmers. It also builds on social capital which involves people's capacities to work together to solve common problems.

Adopting a sound sociopolitical strategy in farming

Good governance, the rule of law, transparency, the elimination of corruption, sound public administration and respect and protection for human rights are useful in setting a sound socio-political environment for sustainable agriculture at the subsistence level. Democratic governments are more likely to be responsive to the needs of all their citizens.

Despite serious soil and climate constraints, low farm productivity and the persistence of hunger and food insecurity in sub-Saharan Africa must at some point be understood as a failure of governance at the national level. Many governments in the region fail to provide essential public goods such as civil peace, rule of law, and investments in public infrastructure, essential services and research. Kaufmann et al (1999), reported that good governance in Africa scored very low using an aggregate governance indicator technique. Poverty reduction through agricultural productivity growth is difficult in Africa and some developing countries partly due to the failure of so many governments in the region to preserve internal peace. Violent conflicts reduce agricultural productivity even at the subsistence level thus compromising secure access to food in multiple obvious ways. In rural farming communities the recruitment of able-bodied young men into army and militia first takes away labour from food production thereby reducing rural incomes. These armed conflicts deter rural subsistence farmers from food production, this could be the reason why countries experiencing conflicts record a significant drop in food production. Also violent internal conflicts encourage emigration of citizens of such area who become refugees in another country thus depending on food aids for survival. Good governance therefore must be embraced by developing economies to ensure peaceful co-existence in rural communities in order to foster subsistence farming.

Government should develop a sound policy that will encourage the establishment of rural infrastructure. The deep poverty of rural areas in developing countries tends to be associated with inadequacy or lack of infrastructure to ensure access to markets and fundamental public services including water and sanitation, education, public health, electricity and good communication network. These social amenities must be provided in the rural communities to boost subsistence agriculture and reduce rural-urban migration of rural labour-force. Inadequate public investments in rural transport infrastructure have been critical to farm productivity growth in many developing countries. Africa's poorly developed rural transport infrastructure is particularly damaging to the interests of rural women, since they bear the transport burden more than men. Throughout sub-Saharan Africa, women contribute at least 65% of household time spent on travel and transport. Rural roads not only open new areas for production, they also increase labour efficiency (Craig et al, 1997).

A farm productivity revolution will be difficult to launch in Africa as long as transport costs remain high due to underdeveloped rural transport infrastructure. High cost of transport impedes farmers from procuring needed inputs and can also reduce the ability to sell their produce for higher prices in cities. Marketing costs for agricultural products are higher in Africa than anywhere in the world and this hurts the rural subsistence, farmer who are sellers and buyers of food. There is an urgent need for a rural broad based investment in transport network including feeder roads to link up most farming communities in the developing countries. Governments need to develop a rural transport policy to address the issue of increased food production in the rural areas. Governments in the developing countries need to develop a strategic framework aimed at revitalising agriculture through appropriate formul ation and implementation of good policies in the sector. This is because agriculture is viewed broadly as a dynamic sector interconnected to the rest of the economy in addition to its role of creating economic growth by generating jobs, rural and urban incomes, savings, poverty reduction, enhancement of the natural resource base and fostering greater social contributions, including domestic tranquility. The overall agricultural environment must be conducive to the changing requirements of producers and rural residents as they respond to the needs of increasingly distant consumers and competitive producers and agribusiness. Government also needs to develop a dynamic market system and complementary infrastructure services in the rural areas to boost subsistence agriculture. The era of "produce and sell" has ended. Knowledge of consumer needs and product promotion has become paramount for linking local capacities with national, regional and international needs. Rapid improvements in farm-to-market roads, regional packaging points, rail and port facilities, refrigeration facilities, accesses to modern telecommunications facilities and timely product and price information for the rural subsistence farmer are essential (Carney, 1997).

Government should create access to knowledge about production and processing technologies relevant to local conditions and changing market opportunities. A new government approach should focus on maximising economically productive opportunities in targeted rural communities. The new policy environment should aim at reversing the negative terms of trade towards the rural sector, stimulating demand for local products and creating a more favourable environment for local, national and foreign investors. Targeted and co-ordinated activities by producers and investors would result in enterprises that will generate farm and non-farm employment opportunities. This will expand local market opportunities and create sustainable economic growth (Bendavid, 1989; USAID, 1994).

Private and public sector investments can be facilitated by targeted incentives and activities that support selected infrastructure, education and training programs and basic health services. Education and health needs are particularly critical and have been a major factor in the traditional low productivity levels associated with rural residents (Schuh and Jungito, 1993). Given the rapidity of change and the legacy of centre-based government programs, the need for extensive participatory effort between government, rural residents and the private sector need to be stressed along with decentralised operations and promotion of community based organisations (CBOs).

Women empowerment strategy

Women play important roles in food production, management of natural resources, income generation and caretakers of household food and nutrition security. Research shows that giving women the same access to physical and human resources as their male counterpartners increases agricultural productivity dramatically. In many countries increasing the assets that women control has a positive impact on the next generation particularly on their education and health. Gender equality is important in subsistence agricultural development. Women need access to productive asset, education, extension services and information, organisations and social networks, credit, legal rights and a voice in the political system. Economic, political, social and cultural institutions and practices that prevent women from reaching their full potential should be reformed. Egalitarian laws and norms work best when both men and women have equal economic opportunities. The human rights of women must be on a par with the rights of men. Legal, political, social and cultural institutions must be changed to create an environment in which women can realise their full potential and contribute effectively to the development of rural subsistence agriculture. Social entitlement and transfer programs must be devised to benefit women, and property rights laws including land ownership rights must be reformed to allow women to hold individual or joint titles to land.

Women are responsible for generating food security for their families in many developing countries particularly in sub-Saharan Africa. Women not only process, purchase and prepare food, but also play a significant role in national agricultural production, producing both food and cash crops. Women account for 70-80% of household food production in sub-Saharan Africa, 65% in Asia and 45% in Latin America and Caribbean. They achieve this despite unequal access to land, inputs such as improved seeds, fertiliser, and to information. Women farmers have shown that they can be experts in their own domain. Women are major contributors to agriculture and play a prime role in ensuring the food security and nutrition status of their household members, yet women could achieve much more in food production, provision and utilisation if agricultural researchers, plant scientists, extension agents and policy makers would level the agricultural playing field. Women face an uphill struggle as a result of weak land tenure rights, exclusion from active role in seed development and selection, neglect by agricultural extension services and barriers to access to complimentary inputs such as fertiliser improved seeds and credit. These impediments are likely to result in forgone economic growth through lower crop yields, delayed adoption of new technology and plant varieties and environmental degradation. The women empowerment strategy (WES) will ensure active participation of women in subsistence agricultural development if fully implemented by the various nati'm Uonal governments of the developing countries.

Conclusions

Strategies aimed at fostering subsistence agriculture must recognise the farmers' cultural, socio-political and economic background. The research strategy will only be successful if it starts with the farmer in terms of project initiation, formulation and implementation. All technologies developed must be adapted to specific locations, for them to be useful. Farmers indigenous knowledge must be recognised, analysed and improved to enhance agricultural sustainability, environmental safety and conservation of agrobiodiversity. Subsistence farmers must be empowered economically by giving them access to credit and other improved inputs. Good governance must ensure proper infrastructural development in the rural areas, improved marketing opportunities, good road network, peace, and equal land rights to women farmers. Women should be empowered economically, socially, and politically to enhance agricultural productivity beyond subsistence level.

References

Benavid, A. 1989. "Rural-urban Linkage Farming and Farm Households in Regional and Town Economics." *Review of Urban and Regional Development Studies* 2: 87-97.

Beyer, J. L. 1980 . "Africa." *G. A. Klee (ed) World Systems of Traditional Resource Management*. London.

Carney, D. (ed). 1998. *Sustainable Rural Livelihoods: What Contributions can We Make? London.* Department for International Development (DFID).

Carney, D. 1997. *The Role of the Department of International Development in Agricultural Development London.* Overseas Development Institute.

Chambers, R. 1980. "The Small Farmer is a Professional." *Ceres* 19-23. Craig, Barbara, J. Pardey, and Johnannes Roseboom. 1997. "International Productivity Patterns: Accounting for Input Quality, Infrastructure and Research." *American Journal of Agricultural Economics* 79(44): 1064-1076. November.

Egger, K. 1979. "Ökologie als produktive Kraft. Erfahrungen beim "Ecofarming" in Ostafrika." Elsenhans, H. (ed) *Agrarreform in der dritten Welt*. pp 217-254. Frankfurt/New York.

Hartmann, M. 1973. Die methodischen Grundlagen der Biologie. *Ann philos* 11: 235-261. Berlin.

Hatch, J. 1976. "The Corn Farmers of Motupe: A Study of Traditional Farming Practices in Northern Coastal Peru." *Land Tenure Centre Monograph I.* University of Wisconsin, Madison.

Hazell, P. and T. Reardon. 1998. "Interaction Among the Rural Non-farm Economy, Poverty, and the Environment in Resource-poor Areas." Paper Presented at an IFPRI/World Bank Sponsored Workshop on Strategies for Stimulating Growth of the Rural Nonfarm Economy in Developing Countries. Warrenton, Va., U.S.A.

Hazell, P., P. Jagger, and A. Knox. 2000. "Technology, Natural Resources Management and the Poor. Background Paper Prepared for International Fund for Agric Development's Rural Poverty Report 2000." *Report of IFAD's Workshop on Rural Poverty.* Rome, Jan. 24-25.

IFAD (International Fund for Agriculture). 2001. *Rural Poverty Report 2001: The Challenge of Ending Rural Poverty.* Oxford. Oxford University Press.

Jazairy, I. M. Alamgir and T. Panuccio. 1992. *The State of World Rural Poverty: An Inquiry into its Causes and Consequences*. Rome. International Fund for Agric Development.

Kaufmann, Daniel, Aart Kraay, and Pablo Zoidolobaton. 1999. "Aggregating Governance Indicators." *Policy Research Working Paper* 2195. Oct. Washington, DC.

Kotschi. 1981. " kologischer Landbau als ein Instrument landwirtschaftlicher Entwicklung." *Entwicklung und ländlicher Raum* 15(5): 7-10.

Pinstrup-Anderson, P., R. Pandyalorch, and M. Rosegrant. 1997. "The World Food Situation: Recent Developments, Emerging Issues and Long Term Prospects." *2020 Vision, Food Policy Report.* Washington, DC. IFPRI.

Reardon T., K. Stamoulis, M. E. Cruz., A. Balisacan, J. Berdugue, and K. Savadogo. 1998. "Diversification of Household Incomes into Nonfarm Sources. Patterns, Determination and Effects." Paper Presented at an IFPRI/World Bank Sponsored Work on Strategies for Stimulating Growth of the Rural Non-farm Economy in Developing Countries. Warrenton, Va., USA.

Schuh, E. G. and R. Jungito. 1993. "Trade and Development in the 1980s and the Challenges for the 1990s." *Latin America Agricultural Economics* 8 (June): 377-400.

USAID (U.S. Agency for International Development). 1994. *Making Markets Work for the Rural Poor: An Agenda to Advance Broadly Based Sustainable Rural Economic Growth in Latin America and the Caribbean.* Washington, DC.

World Bank. 2000. *World Development Report 2000/2001: Attacking Poverty.* Washington, DC.

STRATEGIES FOR FOSTERING SUBSISTENCE AGRICULTURE

Gabriel O. Ihejirika

Abstract

Subsistence agriculture is a system in which a farmer cultivates crops and rears animals in order to produce food for use by himself and his family only. The early man lived by hunting wild animals and gathering wild fruits. He used primitive tools like stones, bows and arrows, traps, clubs and ploughs and there was little to eat and no reserves were made for the next farming season. Fostering subsistence agriculture involves establishing programmes, which sought to improve the incomes and welfare of poor rural women or households through more productive, participatory, sustainable and equitable management of land, water, and forest resources.

Common to these programmes should be efforts to give rural women households secure access to natural resources: to help the responsible government units such as crop departments, animal departments, irrigation agencies and fisheries bureaus redefine their role as facilitators of development rather than the 'doers' of development: and to empower local communities to play a more effective role in the resource management and women produce at subsistence level.

Funding for agricultural research centres will help to produce new varieties of basic crops and animals and when combined with water, fertiliser, pesticides, disease control chemicals, higher prices, and effective extension services will increase yields.

Majority of our rural women or households live in remote areas of high ethnicism. The socio-economic and biophysical conditions in which they live are highly diverse, complex and constantly changing. There was a substantial contribution of women to subsistence agriculture and household food consumption.

The strategies should be to start with people, not the resource and target should be on specific groups of low-income households and not the aggregates of rural poor in a given country. It should start with a problem, not the solution so as to deal with an openness to whatever factors that may turn to be relevant. It should reflect the diverse, complex and dynamic nature of rural conditions.

Introduction

Food disappearance in the world should be of high concern to a patriot. It is difficult for food to be available in quantity and quality to feed the teeming population. In spite of the fact that there was a quantum jump in food importation, the country has hardly ever provided enough food for its teeming. For instance, between 1970 to 1979, the average annual deficits in per capita calorie intake was 24.4 % and between 1980 and 1989 and 1990 – 1996. It became 28.58 % and 8.28 % respectively (ADB, 1996).

This situation does not favour food security, since the majority of the population cannot have access to enough food for active health life at all times of the year (World Bank, 1938). The minimum per capita daily calorie intake recommended by FAO and WHO for maintaining the human body is 2500 Kcal/day (FAO, 1995; World Bank, 1994).

The history and development of agriculture is closely related to the history of man and his search for food. The early man lived by hunting wild animals and gathering wild fruits. The supply of nutrients such as protein was no great problem. He was a hunter and animals provided him with the protein he needs but as the population increased and as man changed from a nomadic hunting system to a settled agricultural one, it became necessary for plants to furnish some proportion of his protein intake. The primitive man uses primitive tools like stones, bows and arrows, traps, clubs and ploughs. There was little to eat and no reserves were made for the next farming season (Tindall, 1975).

Subsistence farming involves cultivation of crops and rearing of animals to produce food for use by the household only. Here, labour is provided by the farmer and his family, small area of land is used; crude implement like basket hoe, cutlass digger, etc., are used; the produces from the farm are not sold; little capital is required and the yield is also very low.

Objective

The objective of this research is to determine strategies for fostering subsistence agriculture with reference to the role of rural women in agriculture.

Problems of subsistence farming

Subsistence agriculture has problems such as the difficulty in controlling pests and diseases. Though the produce are not for sale, farmers are illiterate and crude tools are used. Labour supply is erratic or unreliable. Also the practice of subsistence farming wastes land as there is no means of replenishing soil, hence the farmer will migrate to cultivate fresh land. Also this process bores down farmers due to the unchallenging of the work; and there is no profit gained, because the produces are not for sale. Subsistence agriculture is highly traditional and superstitious in nature. There is also low capital requirement, which can be provided by friends, money lenders and family members. There is no record keeping done and so no improvement is expected and it is not good for permanent crop cultivation, because the time before maturity is too long.

An advance to subsistence agriculture resulted in cash crop production which is a more advanced stage of agricultural production. The farmer specialises in the cultivation of cash crops like cocoa, kola, rubbers, citrus, groundnuts, oil palm, etc. These crops are for export and they require processing into finished products before they can be consumed and the farmer sells the product in a semi-processed form (Ogieva, 1975). Farmers are vital to the economics of developing countries, yet the yields they produce are low.

Subsistence farmers are unable to invest in machinery, seeds and other inputs, or to use intensive farming methods. They suffer from a lack of land with associated pest an disease management problems such as the inability to practice crop rotation and intense inoculum pressure at the margins of small fields. Also even farmers of cash crops have little influence on market prices.

There is a need for crop protection measures which are cheap, simple, cost-effective and sustainable. Appropriate strategies include the use of resistant cultivars, pest forecasting, biological control, pest and disease exclusion through quarantine, and cultural and chemical control.

Moreso, the problems facing subsistence farmers or small scale or resource-poor farmers in the world today are complex.

- Subsistence farmers have no money to invest in modern, cost-effective technologies for crop production. Example crop protection equipment like spray pump may not be within their scope and improved seeds must have great advantage before they can sacrifice to buy them.
- Subsistence farmers have very little, in most cases less than 5 hectares and he/she has to subsist on it. The farmers may not be in a position to adopt crop rotation if they are also to feed their families.
- From crop protection perspective, subsistence farmers have some unique problems because of their small fields. Example, the inoculum from neighbouring crops is often abundant and there is generally much spread of inoculum between fields.
- Due to limited labour and funds, they cannot exploit intensive farming methods.

- The governments of subsistence farmers have attempted to help them by introducing mechanisation on a communal basis. This has not worked well, basically due to management problems.
- Mechanisation, if it is to work properly, should be used at almost all stages in the production process. This is not the case of subsistence agriculture. Example, they may get a tractor to open up a large piece of land and yet have to weed by hand and use human labour for crop protection activities.
- In the case of cash crops, the subsistence farmers play no role in pricing their commodities, both within and in particular outside their countries, which perpetuates their poverty.

The subsistence agriculture therefore require crop protection measures that are:

- Cheap, so that they can afford them.
- Simple, so that they can be applied under particular circumstances.
- Cost effective, so that they can enable farmers to make profits and prosper.
- Sustainable, so that the production is also sustained.

Environment and biodiversity

This world is facing two serious crises: the population crises, and the environmental crises. Between these two, the environmental crises is more serious (Noibi, 1990). Bunnet (1973) observed the exploitation of the natural environment either in form of deforestation or mining of solid minerals and crude oil, and agro-infrastructural developments has resulted in the creation of numerous problems, which has adversely affected the natural ecosystem (plant and animal communities) of many part of the world..

Biodiversity is considered at three levels: genetic, species and ecosystem. Genetic diversity is the total genetic information contained in the genes of individual organism; species diversity refers to the variety of living organism. Ecosystem diversity relates to the numerous diversity of habitats and biotic committees, as well as to the varieties of ecological processes within ecosystems (McNeely et al, 1990).

Strategies for fostering subsistence agriculture should be by increasing the incomes and livelihoods of disadvantaged rural households by devising approaches to resource management that are more productive, stable, sustainable and equitable reputable. This broadly stated, as applied to specific people resource relations, necessitated a comprehensive approach.

Since the aim was to make a significant difference, local experimentation had to be linked to national policy, research, experimentation, and training had to be integrated. And collaboration among public agencies, research and training institutions, and NGOs proved to be essential (Geithner, 1998).

It is worthy to note that we scientists have often been wrong concerning tropical agriculture, and it is important that we recognise our limitations. We have more knowledge about microscopic matters, but local farmers probably know more where continued observation and knowledge of inter-relationships are involved. Green revolution agriculture has been effective mostly in flats irrigated lands where farmers can control the environment. In contrast, most rainfed tropical agriculture occurs in complex, diverse and risk-prone environments. Research stations even in such areas tend to be resource-rich, and have led to only incremental gains for resource-poor farmers. Subsistence farmers develop diverse cropping system in response to an uncertain environment, and our discipline oriented training neglects their complex linkages. Subsistence farmers should be provided with information allowing them to make the decisions and analysis, and accepting that they are fully capable.

Role of women in agriculture

Women account for 70-80% of household food production in sub-Sahara Africa; 65% in Asia and 45% in Latin America and Caribbean (Qusumbing, 1996). They do this despite unequal access to land and other agricultural inputs when compared with men. Poor access to farm inputs is usually a major constraint to women's productivity in agriculture: hence there is need for action to remove obstacles in the path of women.

In the past agricultural development programme has been against women (Afonja,1986). In many cases women do not own land and have no access to fertiliser or chemical and they rely on traditional storage and processing methods. In view of women increasing role in food production (Dickson, 2003), economic policies are now paying special attention at the household level and their consequences for household food security. The reason is that agriculture is needed to foster development in Africa and the role of women as agricultural producers cannot be ignored in economic policies. When women are given equal access to land, resources, human capital and information, women farmers can achieve yields equal to or even, as some studies show, significantly higher than those of men (Cleaver and Gotz, 1993; Cleaver, 1993).

The weakness of women's access to land and other agricultural inputs can result in their inability to possess collateral to obtain credit. Poverty and gender inequality pose a great threat, because of the nutritional benefits associated with increasing women's income (Aboaba, 2000).

Nutrition security which is adequacy in energy, protein, micro nutrients and minerals for all household members, depend not only on sufficient food but also on factors such as health, access to care and clean water is almost the exclusive domain of women. There is a need to take stock of the Nigerian women to identify obstacles in the path of women's productive role to household food security and nutrition.

Women involvement in non-timber forest production

Non-timber forest products (NTFPs) are a particularly important part of multiple-use strategies, because they increase the range of income-generating options for forest-dependent villages while avoiding some of the ecological costs of timber cutting. NTFPs include a range of products with both subsistence and market value, such as foods, folder, gums and resins, oil seeds, medicinal plants and building and crafts materials. NTFP-based activities often are small in size, managed by households and by women involve a diversity of products, provide supplementary income during lean seasons, are labour intensive, use simple technologies, and are accessible to low income and socially disadvantaged groups. The result of a study carried out in west Bengal show the importance of non-timber forest products to women.

a. Three times as may women as men are involved in the gathering of NTFPs.
b. Consumption of NTFPs is equal by women and men.
c. Processing of NTFPs is exclusive to women.
d. Twice as many women as men are involved in the marketing of NTFPs.
e. Seventy-one species are collected exclusively by women, 23 exclusively by men, and 10 species by both.
f. NTFPs account for 20% of household income.
g. Women are mainly responsible for the manufacture of plates made from sal leaves.
h. Women are mainly responsible for about 75% of the marketing of mushrooms, fruits, mahua flowers and liquor.
i. Men are mainly responsible for marketing the leaf plates, kendu leaves and mats (Campbell, 1992).

Results and discussions

Investigation revealed that in spite of the improvement in access to land, the size of individual farm was small thus, women produced only at subsistence level specialising in production of food crops such as yam, cocoyam, groundnut and vegetables and only about 10% of female farmers planted improved varieties.

This has a lot of negative implication for the quality and quantity of food available in the household. The level of food production by this women cannot transform them economically, neither is it enough to provide the household with the highly nutritious food items needed to maintain the adequate health and normal status of the family.

The contribution of women to household food consumption in four zones of Nigeria was substantial (Table 1), thus, supporting the deduction of the significant role of women in food security.

Table I
Major occupation of women in Nigeria

Occupation	ZONES			
	South-East	South-West	North-East	North-West
Farming and food	82.3	13.6	55.0	42.5
Processing full-time wife (household task)	6.2	8.4	13.1	27.5
Non-farming occupation	6.8	56.3	28.4	24.7
Wage occupation	4.6	56.3	28.4	24.7
Others	0.0	0.7	0.4	1.4
N	1504	2413	1500	1828

Sources: Primary data from UNICEF/FGN, 1994.

Through history, Nigerian women have been impaired by gender problems seemingly made most manifest in cultural and religions affairs. Most cultural and religions do not place them equal with their male counterpart, while some cultures have been quite unkind to them. Example among many traditional communities, the death of a woman's husband often marks the beginning of her suffering and sometimes humiliating treatment. In some families, the husband death denies her title to her late husband's property in which the brothers-in-law are most often very much interested in (Dickson, 2003).

Table II

Proportion of food contribution by women to their households

Proportion of food contributed (%)

Ages	ZONES			
	South-East	South-West	North-East	North-East
<25	24.2	11.6	9.1	18.4
25	20.9	36.6	60.2	60.1
26-50	32.6	40.6	9.5	3.6
76-100	4.2	4.7	1.5	2.6
N	1399	2086	1297	917

Sources: Primary data from UNICEF/FGN, 1994

The Christian religion that seems to be one of the kindest to women also has some cautious words for women. Example, they are admonished to be silent in the church and where they have problems they should ask their husband at home. But most Christians do not believe that this injunction is still meant to be observed today. Not when many women are now very active in evangelism and other areas in the church (Dickson, 2003).

As environmental and health hazards mount as a result of heavy pesticide usage, they lead to deterioration in rural economies and bring forth new holistic perspective energies in food production. This involves dynamically evolving system in which widely divergent agricultural practices and conditions are evaluated, modified and verified so as to create a productive and continuing sustainable agriculture. The concept is environmentally oriented, although people and technology are also given significant weight. The system uses components of related disciplines and concepts such as IPM, farming systems approaches in production, research and extension, agro-forestry and agro-ecology (Dar, 1999; Gips, 1990).

Conclusions and recommendations

Women play significant role in subsistence agriculture and household food security in spite of poor distribution of resources, which left them at a disadvantage in agricultural enterprises. Programmes that should make rural women and households secure access to natural resources, improve incomes and welfare of poor rural women through more productive, participatory, sustainable and equitable management of land, water and forest reserves will foster subsistence agriculture.

It is recommended that:

i. Ownership rights of women on land be promoted.

ii. Burden of indoor responsibilities be lessened to encourage higher participation of women in food production.
iii. Informal education on nutrition, child rearing, family planning and general health cares should be strengthened for rural women.
iv. Research for viable technology, training and education should be site-specific, community relevant and appropriate to farmers circumstances. Researchers must be geared to an agricultural production system that promotes health, economic production and environmental protection.
v. Inappropriate agricultural, economic and trade policies of industrialised countries are obstacles to sustainable agriculture which have trapped both their farmers and those of the developing countries, eroding the resource base on which food production depends.
vi. International organisations for instance on food aid, should instead shift the emphasis to the production of food commodities by the resource poor farmers in food deficit countries.

References

Aboaba, I. O. 2000. "Targeting Nigerian Women for Income Generation: in the Context of Household Economy and Nutritional Status. *Journal of Home Economics Research.*

Afonja, S. 1986. *Land Control and Class in Africa.* Edited by Claire Robertson, New York, Africana Publishing Company.

African Development Bank: ADB. 1996. Selected Statistics on Regional Member Countries.

Bunnet, R. B. 1973. *General Geography Diagram*, Longman, London.

Campbell, J.Y. 1992. "Women Involvement in Non-timber Forest Products in West Bengal. Joint Forest Management, Concept and Opportunities. Proceedings of the National Workshop at Surikund" August 1992. *Societies of Promotion of Wastelands Development* 27-29. New Delhi.

Cleaver, K. 1993. Reversing the Spiral. The Population Agriculture and the Environment. Nexus in sub-Sahara Africa.

——— and A. Gotz. 1993. Reversing the Spiral - The Population, Agriculture and the Environment. Nexus in sub-Saharan Africa.

Dickson, A. O. 2003. *Women in Nation Building. Nigeria, a Nation of Surprises.* Christian Publishing Company Ltd. 80-80.

Dar, W. D. 1990. Basic Elements of the Farming Systems Approach to Research and Extension. Dilman, Quezon City, Philippines: Department of Agriculture, Bureau of Agricultural Research.

Gips, T. 1990. *Breaking the Pesticides Habit: Alternatives to 12 Hazardous Pesticides*. Penang, Malaysia: International Organisation of Consumers. Unions.

Food and Agricultural Organisation. FAO. 1995. "Social Security and Food Security. Who can Afford to Eat?" CERES. *The FAO Review* 152: 31 - 33.

Geithner, P. F. 1998. Forestry for Sustainable Rural Development. *A Review of Food Foundation-supported Community. Forestry Programs in Asia* 1-10.

McNeely, J. A., K. R. Miller, W. V. Raid, R. A. Mittmeier, and T. B. Warner. 1990. Conserving the Worlds' Biological Diversity. Washington DC. and Gland, Switzerland: WRI: World Conservation International.

Noibi, A. S. 1990. "Challenges of Environmental Education in Nigerian Schools". M. B. Lawal and A. A. Mohammed, (eds). *Environmental Education Workshops and Seminar Proceedings*. Lagos. Redfield Nigeria Ltd.

Ogieva, E. B. 1995. *History of Agricultural Development. Comprehensive Agricultural Science.* A Johnson Publishers Lagos, Nigeria. 16-19.

Okolo, D. A. 2003. "Women is Nation Building". *Nigeria a Nation of Surprises.* Christian Publishing Co. Nigeria. 80-85.

Qusumbing, A. R. 1996. "Women. The Key to Food Security. *Food and Nutrition Bulletin* 17(1):

Tindall, H. D. 1975. "Traditional Patterns of Vegetable Production." Commercial Vegetable Growing. Oxford Tropical Handbook Ely House. London. 1-7.

UNICEFIFGN. 1994. *The Nutritional Status of Women and Children in Nigeria. Federal.* Office of Statistics Abuja.

World Bank. 1988. *Poverty and Hunger: Issues and Options for Food Security in Developing Counties.* Washington DC.

——— 1994. "Infrastructure for Development." *World Development Report.* Washington DC.

TOWARDS FOSTERING SUBSISTENCE AGRICULTURE IN SUB-SAHARAN AFRICA: CONSTRAINTS AND TRATEGIES

Magnus Chima Ofoh

Introduction

It has been documented that sub-Saharan Africa has an annual population growth rate of 2.5% but only about 1.5% agricultural growth (IBRD, 1989). About 90% of the production comes from the subsistence farmers who generally crop less than 2ha using indigenous knowledge and technologies. Many countries in this region have made various efforts to tackle the problem of food shortage but the problem is still far from being solved. In Nigeria, for example, various administrations had, since the early seventies, tried large scale mechanised farming using the slogans such as Operation Feed the Nation, Green Revolution, Back To Land, etc., but to no avail. Also several agricultural establishments were set up such as River Basins Development Authorities (RBDA), National Accelerated Food Production Project (NAFPP), National Agricultural Land Development Authority (NALDA), Agricultural Development Authority (ADP), National Agricultural Research Programme (NARP), etc.

Unfortunately, most of the above agricultural outfits further compounded the problem of food insufficiency by cultivating marginal lands using environmentally unfriendly inputs and management technologies such as bulldozers (for massive land clearing) and heavy doses of agrochemicals (inorganic fertilisers and pesticides) leading to environmental degradation. Even in the area of research, the recommendations/new technologies emanating from the agricultural research centres were hardly adopted by the farmers, because they could not adapt to the farmers' environments and circumstances. As a consequence, most countries in the sub-Saharan Africa are not only battling with agricultural stagnation and population explosion but also the most recent problem of environmental degradation.

Considering the vulnerability of this agro-ecosystem, there is need to review the existing agricultural production systems vis-à-vis the traditional system (subsistence agriculture). Though the subsistence agriculture may have the problem of not meeting up with the food requirement of the teeming population and the livelihood of the farmers, it has proved to be ecologically friendly. The productivity of this system, however, can be enhanced by identifying the constraints and adopting strategies to overcome them.

Ecological concept of agriculture

An ecological system (eco-system) has been defined as any unit where the organisms interact with the physical environment so that a flow of energy leads to clearly defined tropic structure, biotic diversity and material cycles within the system (Odum, 1971). The components of a natural ecosystem are vegetation, climate, soil and fauna and they exist in a dynamic equilibrium. The dynamic nature of the system accommodates moderate changes, e.g. evolution of new species) or perturbances (such as drought or flash fire) without disastrous consequences. An agricultural ecological system (agro-ecosystem) can be regarded as an ecosystem whose structure and function have been modified by

S/ N	Characteristics	Natural ecosystem	Agro-ecosystem
a.	Biomass supported/unit energy flow (B/E)	High	Low
b.	Net community production (yield)	Low	High
c.	Total organic matter	High	Low
d.	Diversity and complexity	High	Low
e.	Inorganic nutrients	ntra-biotic	Extra-biotic
f.	Nutrient turn over rate	Slow	Rapid
g.	Nutrient cycle	Closed	Open
h.	Role of debritus in nutrient cycles	Important	Unimportant
j.	Stability	High	Low

Table 1: Comparison of some characteristics of natural ecosystem and agro-ecosystem (Modified from Odum, 1971)

man to produce food, fibre or other products. The components of a typical agro-ecosystem (farm environment) are as follows:

a. Physico-chemical e.g. climate, soils, etc.
b. Biological e.g. crop, animal, pests and diseases, soil organic matter, soil living organisms, etc.
c. Socio-economic e.g. labour, capital, market system, culture, etc.
d. External influences e.g. inputs/management technologies.
e. System responses (output).

f. Resource use and conservation.

The above factors are also the determinants of a farming system. An undisturbed natural ecosystem has optimum resource availability and tends to be stable. The basis for the stability can be appreciated when the characteristics of a natural ecosystem is compared with that of a typical agro-ecosystem (table 1).

Peculiarity of sub-Saharan Africa agro-system (crop environment)

The components of the crop environment include the climatic, the edaphic (soil) and the biotic (living organisms) resources. The climatic and the edaphic components are the major ones that influence crop production in this region. Their features are quite distinct from those of the temperate and are discussed below.

The climatic component include rainfall, temperature, light and air (gases). Rainfall limits productivity in many parts of the region through excesses and deficits often in the same location. When the rain comes, it falls in storms with the result that the crops cannot make use of it. The constant high temperature also increases the evaporation rate thus resulting in high water deficits especially during the dry season. The total amount of solar radiation available to the crops during the rainy season is low, because of cloud cover and this lowers yield. The high rainfall causes leaching and erosion while the high temperate causes rapid decomposition of organic matter thus resulting in poor soil fertility.

The soils (particularly of the humid tropics) are fragile and mostly ultisols, oxisols and alfisols. These soils are grouped as low activity clay soils characterised by low cation exchange capacity, and low moisture holding capacity. The ultisols and oxisols which are very common have very low inherent fertility, multiple nutrient deficiencies, nutrient imbalance, high soil acidity and toxic levels of aluminium and manganese and are prone to erosion (particularly ultisols) on exposed sloping land.

Under the natural conditions, these soils are able to maintain their vegetative cover because of the equilibrium (biocycle) that exists between the forest vegetation and the soil – the nutrients move from the soil to the vegetation and back to the soil through litter fall (Ruthernberg, 1971). Once the vegetation is removed during cultivation, the equilibrium is broken. There is increase in chemical and biological processes due to increase in temperature, moisture and sunlight reaching the soil. The carbonaceous materials and the number of micro-organism are also increased leading to high rate of mineralisation and release of nutrients such as nitrate nitrogen which may be eventually lost by leaching resulting in reduced soil fertility. Also the run-off and reduced infiltration result in soil erosion. The extent of the soil erosion will depend on the size of the land opened up. The fragile nature of the environment demands special handling for our agricultural production to be sustained and this can only be achieved by embracing farming systems that are environmentally friendly.

Farming systems and sustainable agriculture

A farming system or an agricultural production system according to Okigbo (1982) consists of one or more business activities or enterprises in which resources consisting of land, labour, capital and inputs such as fertilisers, machines, etc., are through various practices manipulated to create conditions favourable for the growth and development of selected organisms (crop plants, and/or domesticated animals). He further stated that the farming system adopted by a given farming household results from its members, with their management know-how, allocating the three factors of production (land, labour and capital) to three processes – crops, livestock and off-farm enterprises in a manner which within the knowledge they possess will maximise the attainment of the goals they are striving for. The farming system, therefore, is determined by the environment in which the farmer or farm family operates.

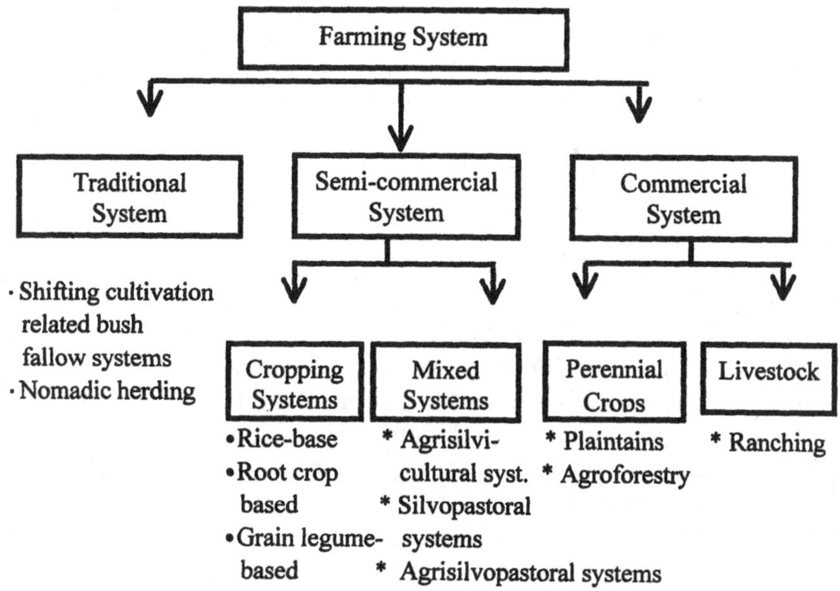

Figure 1: The pre-dominant farming systems (adapted from Lal, 1995).

The major types of farming systems are: traditional and transitional, semi-commercial and commercial systems. The details are shown in figure 1 above.

These systems are based on either extending farmland, intensifying agricultural productivity or by increasing the amount of food crops in relation to cash crops and vice-versa. As the population pressure increased, the traditional farming systems shifted from shifting cultivation (land use factor greater than 10) to intensive agriculture (land use factor less than 2) with high input and management technologies.

The replacement of the traditional bush-fallow system with intensive and mechanised (high input and management technologies) agriculture has adversely affected the sustainability of agriculture in sub-Saharan Africa. The excessive use of fossil-fuel energy and agro-chemicals (such as inorganic fertilisers and pesticides) though had increased yield in a short-run but has the following negative consequences:

- decline in soil productivity due to erosion and accompanying loss of organic matter and plant nutrients;
- pollution of surface waters by agricultural chemicals and sediments;
- increased resistance of weeds and insects to pesticides and feed additives;
- detrimental effects of agricultural chemicals on food quality and,
- increased cost and uncertain availability of energy and chemicals.

An agricultural system that depends mainly on external inputs and impoverishes the environmental quality and the resource base on which agriculture depends is not sustainable.

A sustainable agriculture was defined as that which over a long term enhances environmental quality and the resource base on which agriculture depends, provides for basic human food and fibre needs, is economically viable and enhances the quality of life for farmers and society as a whole (CGIAR, 1988).

Weerakoon and Schall (1989) stated that a sustainable agriculture could be achieved by the following practices:

- providing measures of controlling weed and other pests;
- providing measures of maintaining and improving soil fertility;
- providing measures of controlling erosion;
- providing firewood for cooking and timber;
- providing fodder for livestock production;
- providing a better balanced diet to farmers;
- providing additional cash crops;
- providing shade and wind breaks;
- utilising low cost inputs.

For a farming system to be sustainable in sub-Saharan African agro-ecosystem, it must adopt management practices that simulate the situation in the natural ecosystem. As the traditional agriculture (subsistence farming) has the capacity to satisfy the above condition, it has to be embraced and further developed for increased productivity and sustainability. For the purposes of further discussions, subsistence agriculture will be referred to as the traditional farming system and others regarded as conventional or modern farming systems.

Potentials of subsistence (traditional) agriculture as a sustainable farming system

The basic aim of subsistence agriculture is to produce enough food to feed the family. It is a traditional farming system that tend to simulate the situation in a natural ecosystem thus maintaining the fragile balance between production and conservation. The system is born out of the knowledge derived from the interaction of people with their environment (using their local resources and natural processes, indigenous knowledge and skills, social relations and experiences adopting practices that fit into their local environment).

The farmers clear and till the land manually using machetes, axe, hoes and other indigenous implements which cause the least soil disturbance and minimal problems of run-off and soil erosion. They practice mixed-cropping systems (biodiversity), based on an upper storey of trees and palms that may produce timber, fruits and nuts as well as shade, a middle storey of cocoa, kola, coffee and spice bushes and a ground storey of maize, beans and root crops or other arable crops (Eden, 1980). The stability of this system is derived from crop diversity, which provides a hedge against unexpected climatic changes. The complex above ground plant architecture also allows light to be intercepted at the various levels thus making maximum use of light.

The biodiversity also helps to regenerate the soil by maintaining permanent covers. The covers are the source of organic matter, which helps to stabilize the soil moisture and temperature in the surface layer. The layer therefore becomes a favourable habitat for soil life (plant roots, worms, insects, micro-organism etc.) which uses the organic matter, recycling it into humus and nutrients. The trees and shrubs still function as fallow trees observed in the traditional bush fallow cultivation system. Livestock production is often integrated into the system (mixed farming) so that the prunnings from the trees and shrubs are not only used as mulch and fertiliser for good crops but are fed to the livestock. The wastes from the livestock serve as manure. Subsistence agriculture therefore not only assures yield and environmental stability but also ensures the health of the farmers and consumers by minimising the use of non-renewable inputs (particularly pesticides and inorganic fertilisers).

Constraints of subsistence agriculture and strategies for improvement

Agricultural production in sub-Saharan Africa is generally bedevilled with environmental, infrastructural and economic constraints irrespective of the farming system adopted. For increased productivity to be achieved, the strategies adopted in subsistence farming are quite different from those of the conventional. For example, the traditional system adopts low external input technology that fits into the existing infrastructure and economy (e.g. use of crop varieties resistant to pests and diseases, biological nutrient fixation and recycling etc.), whereas the conventional adopts high external input technology based on improved infrastructure (e.g. high yielding varieties, chemical inputs, machinery and strong extension effort).

From the discussion so far, it can be deduced that the constraints of subsistence agriculture towards increased productivity and sustainability are mainly socio-economic in nature. Some of these constraints are highlighted below and possible strategies suggested for improvement.

Land tenure system and farm size

Many forms of communal, family and compound ownership of land are found in sub-Saharan Africa. They are characterised by rights in land being shared among individuals within the community, family and compound. Inheritance, therefore, involves subdivision of the land over which a man holds user rights amongst his sons. The population determines the size of land available to each individual and it is usually less than 2 hectares.

There is need to encourage individual ownership of the land so that the owner will be able to plan for the agricultural development and be assured of the dividends from his investment. Individual ownership of land will equally encourage the disposal of the land at will and acquisition of more when the need arises. The various governments should reform the land tenure system so that there will be legal ownership of land by individuals or organisations.

Meanwhile, farmers with similar interests should be encouraged to consolidate their land to enjoy the economy of size.

Farm labour

Subsistence farming makes heavy demand on labour (labour intensive). Land is usually cleared and tilled with manually powered tools such as cutlass, axe, hoe, etc. Weeding is done with hoe and harvested grains threshed by beating with stick or trampling by human. The family labour constituted the main source of labour as the subsistence farmer was producing food to feed only his family. This family labour force consisted mostly of members of the immediate/nuclear family but may include extended family members. It is usually dominated by women. According to Synder (1990), women produced 80% of food and other agricultural products in Africa. The level of production did not matter as long as the farmer had sufficient food to feed his family.

With the need to increase food production due to population pressure, the family labour could no longer cope. The farmers now resorted to hired labour. Presently, the labour cost is expensive because of rural-urban migration and competition for labour among industries, infrastructural development and agriculture. Another factor that has raised the labour cost is the new interest in education among the youths.

There is, therefore, the need to introduce some gender sensitive and environmentally friendly labour saving devices for some of the farm operations. The labour-efficient technologies may be based on animal power or moderately motorised and should be affordable by the resource-poor farmers.

Credit facilities and subsidies

When subsistence agriculture was carried out mainly for feeding the farmer and his family, there was not much need for credit facilities and subsidies. The reason was because the farmer was cultivating a small area of land and used simple inexpensive tools. Now that there is population explosion, and need to increase production, there is equally need for credit facilities and subsidies. These facilities will enable the farmer to purchase more planting materials, farm implements, hire labour, acquire and open new land, etc. Usually, the commercial banks hesitate to provide the required credits, because the farmers cannot provide the required collaterals.

The governments of the countries in the sub-Saharan Africa should assist in making these credit facilities available to the farmers by acting as guarantors. Governments could as well undertake to subsidise the prices of the agricultural inputs. It has to be noted that other sources of finance and credit facilities to the farmer are exploitative and unattractive to the farmers.

Transport storage and marketing facilities

The farms are usually situated in inaccessible areas thus making the transportation of the farm produce difficult. As there are no storage facilities and these produce are perishable, farmers are forced to dispose of them in the local markets at very poor prices. The remaining produce unsold is left to waste. This situation dampens the spirit of the farmers and they might not want to increase their production.

There is need to construct feeder roads to enable the evacuation of the produce. At the same time, governments should be involved in purchasing and storing the products. Once a farmer makes profit from the sale of his produce, that is an incentive for increased production.

Research and extension

Until recently, the research and extension work in agriculture were not based on the traditional farmers needs, capabilities and environmental and socio-economic conditions. The research and extension programmes merely concentrated on technological solutions to problems, which are also environmental, cultural, social and economic in nature. Consequently, the farmers did not adopt most of those new technologies. However, this approach to research was replaced by "Farming Systems Approach" to research or Farming Systems Research (FSR). The FSR tried to develop technologies that could be adapted to farmers' needs and production circumstances but this was researcher-managed, thus denying the farmers the opportunity of making use of their indigenous knowledge and skills.

There is, therefore, need to adopt some new approaches to research and extension which are "participatory" in nature. Some examples of such approaches include Farmer Field School (FFS) and Farmer-To-Farmer extension. These approaches have enabled the farmers to analyse their situation and question some anti-ecological technologies. They are then able to select and adapt the technologies most appropriate to their local agro-ecological and economic conditions, knowledge and skill which they communicate to their fellow farmers in farmer-to-farmer extension. Using the above approach the researchers and development workers become only facilitators.

Gender issues/women marginalisation

A considerable part of the world's food (over 50%) is produced by women and they are responsible for about 80 percent of food produced in sub-Saharan Africa (LEISA, 2002). Culturally, women maintain home gardens which are the reservoirs of agro-biodiversity (Oakley, 2004). They usually plant high-densities of under-utilised species which they belief are adapted to their local environment and cultural legacy. This practice not only ensures nourishing diet but assures food security, economic stability and healthy agro-ecosystem. Apart from production, the women provide 65% of the time required for transporting farm produce to market in sub-Saharan Africa. They usually carry an average load of 20 kg on their heads over a distance of between 1.4 and 5.3 km (Spore, 2004).

Unfortunately, despite their contributions to food production and marketing, these women do not have access to, and control over, productive resources and credit. In Kenya for example, 98% of the women work full-time in agriculture but only 5% of these have land ownership. As a result, these women are constrained and find it difficult to improve on their output and income, thus the vicious circle of low productivity, low income and poverty exists. As women are directly involved in agricultural production, processing, storage and marketing, they should be adequately empowered to effectively catch up with these responsibilities. Their needs and priorities should be considered in formulating policies concerning allocation of production resources; credit, research and development of agricultural technology. They should also be encouraged to acquire more education and training. Most importantly their drudgery could be reduced by the provision of some simple labour-saving devices (technologies) and middle-range modes of transport such as hand carts, wheelbarrows, bicycles, etc.

Conclusion

The agro-ecosystem of the sub-Saharan Africa is quite different from those of the temperate areas. In the temperate, successes have been made in increasing food production through high input technologies and mechanisation. Attempts to do so in sub-Saharan African countries have resulted in ecological destabilisation, more food shortages, malnutrition and poverty. The only option for the countries in this region is to re-embrace and reinforce the traditional farming system unique to this environment (Subsistence Agriculture). The productivity of subsistence agriculture can further be enhanced by tackling the infrastructural and other socio-economic constraints surrounding it.

References

CGIAR (Consultative Group on International Agricultural Research). 1998. *Sustainable Agricultural Production. Implication for International Research*, Rome TAC Secretariat, FAO, March, 1988.

Eden, M. J. 1980. "A Traditional Agro-system in the Amazon Region of Columbia." *Tropical Ecology and Development* 509 – 511.

IBRD (International Bank for Reconstruction and Development). 1989. *Sub-Saharan Africa: From Crisis to Sustainable Growth.* World Bank, Washington DC, USA.

Lal, R. 1995. *Sustainable Management of Soil Resources in the Humid Tropics.* United Nations University Press Tokyo, Japan, 146pp.

LEISA. 2002. "The Feminisation of Agriculture", *Women Managing Change. Magazine on Low External Input and Sustainable Agriculture*, December, 2002. 18(4): 4 – 5.

Oakley, E. 2004. "Home Gardens: a Cultural Responsibility". *Valuing Crop Diversity. Magazine on Low External Input and Sustainable Agriculture.* March 2004. 20(1): 22.

Odum, E. P. 1971. *Principles and Concepts Pertaining to the Ecosystem. In: Fundamentals of Ecology.* W. B. Sanders, ed., pp 9 – 36.

Okigbo, B. N. 1982. "Farming Systems Research. An Overview of its Definitions, Concepts and Scope." Paper Presented at the Training Workshop on Farming Systems Research, NIFOR, Benin City, 30th July – 4th August 1982.

Ruthernberg, H., 1971. *Farming Systems in the Tropics.* Oxford Claren Press, England, 313 p.

Spore 2004. "Land Tenure; Property, Security, Equity." A CTA Information for Agricultural Development in ACP Countries, June, 2004. No. 111, pp 4–5.

Synder, M. 1990. "Women: The Key to Ending Hunger." *The Hunger Project Papers*, No. 8, August 1990, p 37.

Weerakoon, W. L. and R. Schall. 1989. "Aims and Scopes of Conservation Farming." GTZ 1989, 9 pp.

STRATEGIES FOR FOSTERING SUBSISTENCE AGRICULTURE

Fayeye Timothy Rotimi

Abstract

Much of Africa, Asia, the Middle East and Latin America remain poor agricultural economies, with a wide spread subsistence-oriented farming. The ecological and productivity sustainability of traditional agriculture is under pressure, therefore the improvement of subsistence agriculture must be seen as a central element in poverty eradication and national development. Improvement strategies should not, however, destroy the unique advantages of the traditional system. It should rather build productivity enhancing improvement into them. There is the need to harness local resources with minimum external inputs. Government policies and projects should meet farmers' aspiration, promote mainstreaming of the marginalized groups, build on female concepts and farming practices and show respect to farmers' cultures and thoughts. Research should focus more on the improvement of local cultivars and the ecological sustainability of farm practices. There is the need for a more effective agri-cultural information service in local languages especially to the marginalized groups, such as women. The need to tackle the HIV/AIDS pandemic to enhance food security is important especially in Africa.

Introduction

Pure subsistence agriculture is relatively rare today, however, subsistence-oriented farming systems are widespread in poor countries of the world. In Ghana, subsistence usually means that families produce between 50 to 90 percent of their food. Much of Africa, Asia, the Middle East and Latin America remain primarily poor agricultural economies. In Romania (Ross and Ross, 2002) nearly half of the populace dwell in rural areas, the majority of which are involved in small-scale and subsistence farming. In the West African states of Burkina Faso and Chad, 80 percent of the work force engage in subsistence farming. In the developed countries of the world, there are only isolated pockets of subsistence farming, e.g. the crofters in the highlands and islands of Scotland. The primary aims in subsistence farming is to provide enough food for the farmers and their families. It is a diversified, mixed cropping system with low external inputs. Farm size is small, works are carried out without effective implements, as such, it is arduous and time consuming. Only basic food stuffs needed for sustenance are grown, this includes rice (South East Asia) and millet (West Africa). Livestock kept are generally indigenous breeds, flock size is small, productivity low and mortality is high. Food supply is seasonal and sometimes erratic. Animals are seldom sold, but in times of cash shortages and unexpected need in the family.

There is an increasing need for a change in outlook by subsistence farmers due to pressure arising from increasing human population, urbanisation and industrialisation. Hooft et al (2003) opined that ecological sustainability and productivity of traditional agriculture is under pressure. Subsistence farming needs to adjust to the changing technologies, economic conditions, habits and people's aspiration: The slash and burn practices of shifting cultivation contributes to the destruction of primary forests and leads to biological and agricultural deserts (Goeltenboth, 1990). The African humid tropical rain-forest is converted into agricultural land far more rapidly than the Latin American or Asian through methods primarily propelled by subsistence agriculture (Jutzi, 1992). Recent reports (SPORE, 2004) showed that nutrient loss from burning of crop residues could reduce yield by 50 percent. The fungus (Fusarium Oxysporum) is a serious constraint to banana cultivation in subsistence agriculture (O' Donned et al, 1998). African subsistence farming is currently faced with soil degradation, loss of soil fertility, frequent droughts, labour shortages; declining yields and general drudgery associated with human-powered agricultural production. Land fragmentation has led to reduction in farm size. In the Philippines farms are usually not more than 1 to 2 hectares. The pressure on farmers to produce food from increasingly smaller farms resulted in the abandonment of fallow periods, acceleration of soil erosion, and reduced productivity of rainfed subsistence economy.

The plan to improve subsistence agriculture should be seen as a central element of poverty eradication and national development in poor countries. If young people must be attracted to farming, then it must be attractive to them both in cost and returns. A young Jamaican recently made this remark: '*My father spent all his life in the country, on a farm... he must have been a fool, because he has nothing to show for it. I 'm not going to do that.*'(SPORE, 2004).

Strategies for fostering subsistence crop production

Governments in poor economies have in time past subscribed to Western style, high external input agriculture. Multinational seed companies are also vanguard of this approach, for the purpose of promoting their sales of agrochemicals and seeds. Farmers often lose their own seeds and find the cost of agrochemical unbearable. What is needed is a sustainable agriculture - sustainable from the economic, social and environmental perspectives. Recently, Chiwona - Karltun et al (2002) reported that farmers in Domasi area of Malawi cannot simply afford the inputs required to grow maize as a result of difficulty in obtaining subsidy on fertiliser and seed. Attempts to modernise basic grain production without changing the production systems had resulted into rural populations becoming debt dependent, accelerated deforestation and land-water resources degradation (Welches and Cherret, 2002). Soil in humid tropics requires maximum biological protection and low external input for its sustained utilisation. Jutzi (1992) reported that continuous mono-cultural based field crop production on humid, tropical soils, with massive external input is linked with considerable ecological risks and constrains. Improvement strategies must not destroy the unique advantages of the traditional systems, it should rather build productivity enhancing improvements into these systems. Approaches that help in maximising present local resources no doubts will, in the future, help in making effective use of external resources. There are those who believe that external input is the only panacea for the improvement of subsistence agriculture in the Sahel area. Brouwer and Voortman (2002) however, pointed out that external inputs are unaffordable to local farmers in this region. Furthermore, reduction of pesticide use has been reported to lessen health problems and medical expenses (Walsum, 2002).

Farmer oriented government policies

We need to recognise the resources we have and start to improve our agriculture with what is already available, with minimum external inputs. Developing countries are not in want of agricultural projects, however projects are often designed without genuine regard to what has gone before. Repetition of errors of past projects often occur. Project objectives must be consistent with farmers' needs. This requires an adequate diagnosis of the real needs of the farmers.

Recently, a project in Mount Cameroon region was reported (Wiesiy, 2002) to fail in meeting the yearning of the parties involved, because the real needs of the women farmers were improved cassava planting materials and planting technology while the project was busy providing schooling on biodiversity conservation and vegetable farming. The experience of Vieira and Wambeke (2002) in El - Salvador showed the need for leaders' and farmers' organisation for knowledge sharing and capacity building. Policy implementation is often implemented in more developed and accessible areas with poor and remote areas remaining marginalized (Ghotze and Ramdas, 2002). Changes in government must not starve important projects of essential funds. There is also the need to limit government involvement to formulation of essential policies and rural transformation projects that will improve accessibility, rural life and productivity. Government should not venture into establishing arable farms, rather it should provide advisory services on productivity enhancing technology, soil conservation, skill development, marketing, storage and processing of agricultural produce.

Research to enhance local farmers productivity

Research should address the need of subsistence farmers for it to be of any use to them. Chiwona - Karltun et al (2002) observed that whereas breeding for reduction/removal of toxin has featured very high on the research agenda for cassava, the Domasi farmers (Malawi) would prefer a bitter Cassava to sweet cultivars, because it is less attractive to thieves and pests and because it reduces the social responsibility of sharing cassava with neighbours. If farmers and scientists work together and learn from each other, it will be easy to increase productivity sustainably within a short time, even without external inputs.

Farming system research should be focused on the evaluation of local knowledge and systems of agriculture and farm practices. This knowledge is scanty in many schools in poor countries. Research should focus on improvement of the existing systems instead of their wholesale replacement. For technical change to be relevant to farmers, it must be based on good understanding of how farmers perceive and manage their farming systems. To gain such knowledge, scientists must be ready to operate outside their conventional mode and maintain a strong working relationship with extension workers (Farrington, 2002). It is the feedback from functional extension that should direct or reshape future research emphasis. Some donor agency should help to finance adaptive research like the one reported by Wallach (1989) in Sudan, where the Ford Foundation gave grant to the Sudanese National Council for Research to sponsor research competition for multidisciplinary field work in the grasslands. Design and fabrication of appropriate land and animal-drawn implements is essential to enable the peasant farmers to work more effectively, with less physical effort and to save the time spent on agricultural tasks. Ross and Ross (2002) reported that the use of newly introduced broad-fork in Romania reduced physical demand of hand tillage and time by 50 percent or more compared with the traditional methods. The use of the low technology row crop cultivator reduced drudgery and led to an increase of 25 percent (or more) in yield.

Research also needs to provide information on the interactions that exists between local and external inputs. For instance, Preston (1990) observed a significant negative interaction between traditionally used compost manure and organic fertiliser in Kandep (Papua New Guinea, PNG) and concluded that the occurrence of consistent compost X fertiliser interactions whether positive or negative would have implications for the development of subsistence agriculture in PNG.

The rapid multiplication of pathogens under tropical conditions is a threat to subsistence crops. Research is needed on cheap control of viral, bacteria and fungi diseases that are plaguing indigenous cultivars. Collaborative effort to control tsetse fly should be intensified, as this would enhance the use of animal drawn implements beyond the tsetse-fly free belt. Anazodo et al (1986) reported that the use of animal power on the farm could advance the growth and efficiency of mixed- farming techniques in many African countries. Recent import of genetically modified crop into Africa (SPORE, 2004) failed to give the desired result of disease resistance. Improving yield and disease resistance of local cultivars may therefore give better results and equally save the cash crunch economy of foreign exchange.

Effective information service

Agriculture in developing countries is carried out by poor mostly uneducated people, a large proportion of them are women and old men. Mastery of native dialect is therefore a prerequisite for effectiveness in information dissemination by extension workers. There is the need to train and retrain local extension agents. Recruitment of extension workers should not be politicised, priority should be given to small, marginalized tribal groups whose basic livelihood is farming and to whom the need for communication in local languages is more crucial.

State-owned agricultural development information net-working services and NGO with interest in the development of subsistence agriculture can do more in getting information on medium scale technology across to farmers through printed and spoken words. Inventions at such centres as National Centre for Agricultural Mechanisation (Nigeria) can be printed in local magazines, newsletter and even regional based magazines and made available to village level extension workers to translate, promote and facilitate the adoption of such medium scale technology by the local farmers. Ground Up, a regional magazine published by Participatory Ecological Land Use Management Association, is been used to disseminate useful information among its members in Eastern and Southern Africa.

Radio has a long track record but it's often overlooked. Wasori (2000) pointed out that it is not only technical obstacles that need to be overcome to facilitate farmers access to information, we need an attitude of participation and openness. Radio service network will facilitate farmers' access to a variety of information. Programs from Farm Radio Network for instance can be relayed in local farm radio programs where such services are not available. This is currently being done in many Latin America and Caribbean countries. Bennett (2002) reported that a community garden in South Africa experimented with a bamboo irrigation system developed in Thailand after hearing about it on a local farm radio program.

Ecologically sustainable agriculture

Ezedinma and Youdeowei (1986) defined agriculture as a biological production process, which depends on the growth and development of selected plants and animals within the local environment.

The phase within the local environment underscores the need for ecological emphasis on agricultural practices. An ecologically viable agriculture is the one which simultaneously increases productivity and only minimally decrease future potential use of the environment. High external input agriculture may increase productivity but it also threatens the long-term health of the human environment through accelerated soil degradation. Soil in the humid tropics and Asian - Pacific regions needs maximum biological protection for their sustainable utilisation. 20% of Africa's vegetated lands are degraded. A study of the 100years of disasters in Tanzania revealed that 38% of disasters were caused by floods (Hatibu, 2002). A collaborative study (FA0, 2001) showed that 42% of land in South Asia is damaged by erosion. In Thailand 34% of total land area is affected by water erosion. It was concluded that excessive use of chemical fertiliser, soil erosion and excessive extraction of ground water are the major cause of land degradation in the Asian-Pacific regions. Schussler (2002) opined that we need an agrarian environment, which guaranteed sustainable means of production in order to preserve the food basis of coming generations, guarantees food sovereignty and strengthens rural communities. Soil management practices need to conserve plants nutrients and not aggravate their depletion. A recent report (SPORE, 2004) showed that maize in Guinea savannah region of northern Nigeria never seems to perform as expected. Instead of the 5 tons expected when using fertilisers, farmers were getting only 1 to 2 tons. Researchers from the International Institute for Tropical Agriculture (IITA) blamed a lack of sulphur could double crop yield.

Indiscriminate use of herbicides could poison the cropping environment. Higgins (2001) reported that the use of Round Up (a herbicide containing glyphosphate) in the war against coca plants (the source of cocaine) has wiped out subsistence crops (such as corn, and yucca), kill birds, mammals and aquatic life in the Amazon ecosystem. Money used to procure herbicide can be used in giving farmers incentives not to continue the production of coca plant.

Goeltenboth (1990) reported a site-stable gardening developed by Wau Ecology Institute. This system was shown to have the potentials for improving the life situation of subsistence farmers and the conservation of the mid-mountain tropical environment of Papua New Guinea. The concept of alley farming was developed at IITA in which legume shrubs are integrated into cropping fields to maintain soil fertility and promote continuous cropping. The Quesungual system developed in Honduras (Welches and Cherrett, 2002) is an adaptation of an indigenous agro-forestry system. It was reported to improve water retention quality of the soil due to better infiltration and increased organic matter content compared with the traditional slash and burn system. The losses of nutrients were 10 times lower, crops were sustained for longer period and grain yield was higher with this system. The system has the potential for integrated livestock production, generation of cash income and enhancement of rural development. A conservation agriculture that was experimented in Guaymango (El Salvador) led to 2 to 3 times increase in yield, 2/3 reductions in labour for sowing and better soil water and fertility (Vieira and Wambake, 2002). The adoption of this ecologically beneficial agriculture can be facilitated via favourable policy of environment, motivation and participation for farmers. Brouwer and Voortman (2002) proposed that poor farmers in the Sahel region may have to tap on options developed in other regions or that which evolves in their own environment. They reported that 1.5 tons per hectare of manure, as evenly as possible, every couple of years could allow more regular fertilisation of larger area of land than the current practice of applying 10 tonnes/hectare, and that sheep manure and urine gives better result than cattle manure in the relatively acidic depressions. The productive use of rainwater can be improved in the semi-arid areas if the Mashamba ya Mbugani (Fields located at the bottom of landscape) currently practised in Tanzania is given a wider publicity.

Respect for farmers culture and thought

Underlying the wide range of receptivity and resistance to change in human societies are culture and thought. Knight (1974) submitted that there is ample evidence to convince one that the traditional farmer is rational within his own cultural milieu and that we can only place our knowledge within the farmers' conception of resources, risks and rationality. The farmer's ethno geography is the link between him, his behaviour and his environment. He needs to be provided with the opportunity to understand innovation within an accustomed framework. Batuuka and Nkanda (2003) reported that many techniques were not adopted by Busoga farmers in Uganda, because they are not part of their farming culture. We may not be able to convince local farmers that an innovation is viable until we gain adequate understanding of his risks and uncertainties, his perception and ability to tolerate them.

Tackling HIV/AIDS pandemic

The world food submit action plan endorsed by 185 nations to lift 20 million people above hunger line every year may not be accomplished in the face of HIV/AIDS which primarily affects those aged between 15 to 45 years. This group forms a great proportion of the farm labour and determines the ability of household to produce sufficient food. The spread of HIV/AIDS has increased the number of single parents and female headed households, this has eroded the ability of such families to feed the healthy and plagued members of the household. Tackling the pandemic of HIV/AIDS is therefore strategic to food security in Africa.

Harnessing the livestock potentials

Potentials for meat production is high in both humid and arid regions of the world. Ehui et al (2000) suggested that investment should focus on making available diseases-resistant breeds such as N'dama cattle. Research is also needed in improving the digestibility of crop residues and improvement of animal healthcare. In the sub-humid zone extension, education and credits are very essential. In the semi-arid zone improved use of crop residues in animal feeding, extension and animal healthcare services are important. Hatibu (2002) submitted that livestock production in the semi-arid and arid areas can be improved substantially through concentration of scarce water as well as provision of supplementary water during critical time. There is the need to give pastoralists property rights to a variety of pastures and water resources and the right to move between those resources (Ehui et al, 2000).

We need to understand the logic behind the 'diversified' system and the 'specialised' system, to be able to proffer solutions that will improve the efficiency of small farmers. The diversified system is used by farmers to counteract risks and optimise opportunities under changing and adverse circumstances. Initial and running costs are minimal. Animals scavenge to supplement the little food supplied by the farmer. Mortality can be as high as 400/o to 800/o as a result of infectious diseases, theft, accidents, predators, internal and external parasites and inbreeding. The best way therefore to optimise the diversified system is to reduce the mortality rate of animals.

Low cost vaccination against diseases such as New Castle, fowl, pox and hog cholera and simple housing to protect the animals especially at night are needed. There is an increasing need for knowledge on how to improve the diversified system. Animal science as presently taught in most tertiary institutions does not take into account the complexity of the context in which small scale rural families rear their livestock.

Livestock projects designed to improve farmers productivity have failed in time past for lack of thorough analysis of the reality of small scale farming families, their cosmo-vision, survival strategies, rationale behind their production system, the role of animals within their reality and the way they perceives new projects. For livestock to succeed, there is need for village level participatory approaches and strategic location of demonstration sites. The involvement of women is necessary to surmount religious and social obstacles. Pastoralists need to be involved in management plans far grazing areas. Non-governmental organisations (NGOs) may be involved in organising livestock farmers into groups for the purpose of training and harnessing funds to improve productivity. Women are major players in livestock production. There is the need for a well-funded, functional livestock extension service. There should be adequate female extension workers, veterinarian and laboratory technicians with good mobility and accommodation to facilitate their activities. Government-information service should spur farmers to increase production not only for cultural and ecological reasons but also for socio-economic advantage. Government would perform better as facilitator rather than being a direct producer of livestock.

Building on female concepts and practices

To expand the commercialisation of subsistence agriculture we must be able to build on female concepts and practices. This is because majority of the world's agricultural producers are women, they produce 80% to 90% of the food in sub-Sahara African states, 50% to 90% in Asia and 50% in Central and Eastern

Europe. Ironically, women often experience the worst deprivation especially among the world's poorest countries. In Kenya 98% of women work full time in the agrarian sector but only 5% have land ownership titles. Tengezer (2003) observed that women lack access to information on traditional land use and management on ground of culture.

Women have comparatively benefited less from agrarian reforms in Latin America due to judicial, structural, ideological, cultural and institutional factors (Schussler, 2002). A sample survey in irrigated area showed that Egyptian's rural women participated little in decision making. Men have almost exclusive authority on matters related to seed, buying and selling of livestock and land, what to grow and where to sell produce. In many cultures there are specific tasks traditionally ascribed to men. In 1989, 24% of landowners in Egypt were women, however credit to women under the Principal Bank for Development and Agricultural Credit (PBDAC) were not more than 12% and 16% for short term and investment loans, respectively.

More priority should be given to women empowerment within the framework of national policy. Past reforms were often gender blind. Policy makers and other stakeholders need to understand women's practices and perception about agriculture including cultural and socio-dimensions. There is the need to sensitise, with special attention to gender, the issue of control and assess to resources, credits and market to enhance women's-contribution to agricultural development. Cosmetic changes in policies and initiatives that fail to address the contradictions of the dominant paradigm of agricultural develop are not what is called for (Pionetti, 2002). Many agricultural policies and projects in Uganda have resulted in unexpected negative consequences for women, and they often fail in other objectives, largely because they did not adequately consider the critical and complex roles that women play in most African agricultural systems (Goldman and Heldenbrand, 2002). The call by Lavia compesina and FIAN for agrarian reform based on human rights and creating agricultural environment which guarantees sustainable means of production need to be heeded by all developing countries. Land reforms should also have a compensatory effect on other forms of social, regional and racial discriminations.

There is the need to train more female extension workers to meet up with extension services to the women farmers especially where socio-cultural barriers have restricted women's access to information. NGOs can also engage women in capacity building. Mainstreaming will facilitate women involvement in participatory approach, this will contribute to women empowerment in the long run. They can also assist in organising women into self-help groups. Walsum (2002) reported that women self help groups often function better than the men's group. Self help groups can also form federations for the purpose of training and information. In Anantapur district (South India) the sharing of information among self help groups led to the adoption of the use of gypsum in improving groundnut production by 500 % in just 5 years (Walsum, 2002).

Conclusion

Farmers in subsistence economies cannot afford the high external inputs agriculture of the advanced world. There is the need to harness local resources and minimise external inputs. Policy reforms should address the peculiar needs of various categories of farmers and give grazing rights to pastoralists. Agriculture programmes should be implemented with farmers' participation, showing due respect to their perception and culture. Research should be focused more on disease control measures, improvement of local breeds and cultivars and medium scale technologies. There is need for improvement in agricultural information services especially to marginalized groups like women. Tackling HIV/AIDS pandemic is also strategic to food security especially in Africa.

References

Anazodo, U.G.N., Ezeike, G.O.I. and E. U. Odigboh. 1986. "Farm Mechanization." *Introduction to Tropical Agriculture*, Longman Group Ltd, pp 88-101.

Batuuka, S. and D. Nkanda. 2003. "Changes and Controversies in Uganda." *Compas Magazine for Endogenous Development* 6:38 - 39.

Bennett, N. 2002 "Connecting Farmers Worldwide Through Radio. Changing Information Flow. *LEISA Magazine*. 18(2): 20 – 21.

Brouwer, J. and R. Voortman. 2002. "Soil and Crop Growth Variability can be an Opportunity: Example from Niger." Creating Living Soil. *LEISA Magazine*, 18 (3): 16 –17.

Chiwona, Karltum, L. Katundu, C. Ngoma, J. Chipungu, F. Mkumbira, J. Simukoko and J. Jiggins. 2002. "Bitter Cassava and Women: an Intriguing Response to Food Security." Women Managing Changes, *LEISA Magazine*. 18 (4):14 -18.

Ehui, S., S. Benin and D. Spencer. 2002. "Development Strategies for West Africa." www.ipri.org12020 'focuslfocus041focus0408.htm.

Ezedinwa, F.O.C. and A. Youdeowi. 1986. *Introduction to Tropical Agriculture*. Longman Group Ltd., pp 6 - 11.

FAO, 2001. FAO Regional Office for Asia and the Pacific, Bangkok, Thailand, (www.fao.or.th).

Farrington J. 2002. Recent and Future Challenges in Agricultural Extension. Changing Information Flow. *LEISA Magazine.* 18(2): 6 - 8.

Ghotze, N. S. and S. Randas. 2002. *Women and Livestock Creating Space and Opportunities.* Women Managing Changes. *LEISA Magazine,* 18(4): 16 - 17.

Goeltenboth, F. 1990. Subsistence Agriculture Improvement Manual for the Humid Tropics. *Tropical Agro Ecology,* No. 4.

Goldman, A. and K. Heldenbrand. 2002. "Gender and Fertility Management in Mbale District, South Eastern Uganda." *African Studies Quarterly. The Online Journal for African Studies.* (1 a 2)

Higgins, M. 2001. "War on Drugs Take Toll on Environment." *Environmental Network,* March 2001. www.forests.org

Hatibu, N. 2002. Rainwater Management: Strategies for Improving Water Availability and Productivity in Semi-Arid and Arid Areas. www.iwmi.cgiar.orglhomelraimvater.htm.

Hooft, K. Vant, C. Reigtjes, Harverkork and W. Hiemstra, eds. 2003. "Controversies in Agriculture", *Compas Magazine for Endogenous Development,* 6:34.

Jutzi, S. C. 1992. The Sustainability of Agricultural Food Commodity Produced in the Humid Tropics of Central and Western Africa. *Tropenlandwirt* (Germany). 93(2): 137- 151.

Knight, C. G. 1974. *Ecology and Change: Rural Modernisation in an African Community.* Academic Press, Inc.

O'Donnell, K., H. C. Kistler, E. Cigelnik, and R. C. Ploetz. 1998. Multiple Evolutionary Origins of the Fungus Causing Panama Disease of Banana: Concordant Evidence from Nuclear and Mitochondrial Gene Genealogies. *Proceedings of the National Academy of Sciences of the United States of America,* March, 1998. Vol. 95 (S): 2044 - 2049.

Pionetti, C. 2002. "Small Change Crops - the Marginalization of Women Farmers' Priorities." Women Managing Change. *LEISA Magazine* 18 (4): 22 – 23.

Preston, S. R. 1990. "Investigation of Composed X Fertiliser Interactions in Sweet Potato Grown on Volcanic Ash Soils in the Highland of Papua New Guinea." *Tropical Agriculture* (Trinidad and Tobago) 67:239- 242.

Ross, A. and J. Ross. 2002. New Tools for Romanian Women Farmers. Women Managing Change. *LEISA Magazine* 18(4): 9.

A REVIEW OF THE POTENTIAL ROLE OF BAMBARA GROUNDNUT *(Vigna subterranea (L) Verdc)* IN SUSTAINING THE LIVELIHOOD SYSTEMS OF PEASANTRY FARMING SYSTEMS IN SUB-SAHARAN AFRICA.

Andrew Malekani

Abstract

Bambara groundnut (Vigna subterrania (L) Verdc), also nicknamed as famine crop, drought relief, insurance crop, poor man's crop and food security crop, is an indigenous grain legume grown mainly by subsistence women farmers in drier parts of sub-Saharan Africa. The crop has advantages over more favoured species in terms of nutritional value and tolerance to environmental conditions. It can yield on poor soils with little rainfall. It is a rich source of protein (16-25%). Generally, it contributes much to the nutrition of most subsistence farmers especially those in drought prone areas in sub-Saharan Africa. Despite its potential as a source of livelihoods for subsistence farmers, only little information is available regarding its economic and ecological potential. This paper reviews some of the potential roles that bambara groundnut has on sustaining livelihoods of peasant farmers in sub-Saharan Africa. An element of some policy implications is outlined. It pulls ideas from various research works that has been done on the crop.

Introduction

Most human food requirements are provided by fewer than 20 crop species. There remains a vast repository of many hundreds of under-utilised species that have been grown locally for centuries and which contribute to food security of the world's poorest people (FAO, 2001). One of such crops in sub-Saharan Africa is bambara groundnut *(Vigna subterranean L. Verdc).*

Bambara groundnut is an indigenous grain legume grown mainly by subsistence women farmers in drier parts of sub-Saharan Africa. The crop has advantages over more favoured species in terms of nutritional value and tolerance to adverse environmental conditions (FAO, 2001). The crop has a number of production advantages in that it can yield on poor soils with little rainfall as well as produce substantial yield under better conditions. For example, it has been reported by many researchers that bambara groundnuts can withstand a wide range of ecological conditions ranging from as low rainfall as 500 mm per year and between 750-900 mm per year (Temu, 1994; Duke *et al.*, 1977). Moreover, it has been reported to produce good yield even when heavy rains occur except during the flowering and maturity periods (Ameyaw and Doku, 1983; Ramolemana *et al,* 1996). In terms of soil pH, it has been found to withstand ranges as low as 4.3 (NAS, 1979). It is this adaptability to a wide range of ecological conditions that makes bambara groundnut fit to the peasant farming systems in sub-Saharan Africa.

Origins and distribution of Bambara groundnut

Bambara groundnut is thought to have originated in West Africa. Hepper's (cited by Doku, 1996) exciting account of the findings of wild bambara groundnut confirmed a view that was generally held that the areas of north Yola in north-east Nigeria and Garoua in northern Cameroon are the areas of the centres of origin. Begemann cited by Doku (1996) made analyses of qualitative and quantitative characters of some 2000 International Institute for Tropical Agriculture accessions collected from West, Central, Eastern and Southern Africa observed some variability of these accessions that decreased sharply as one moved eastward and southward from West Africa, indicating that West Africa is the centre of origin of the crop.

From this centre of origin in West Africa, bambara groundnut has spread throughout tropical Africa, warm temperate Eastern and Southern Africa and to other parts of the world. However, this does not reflect that West Africa could be the only place with genetic diversity of this crop. Doku (1996) has suggested that more explorations are required to complete the picture.

Economic, ecological importance and uses of bambara groundnut

(a) Economic and ecological importance

Bambara groundnut has been labelled as a famine and drought relief, insurance, poor man's and food security crop (Masha, 2001). The main reasons for being given such names is its ability to yield well in less favourable growing conditions particularly under moisture stress or unreliable rainfall as compared to closely related legume species such as cowpeas *(Vigna unguiculatca)* (Rowland, 1993; Linemann and Azam-All, 1993). In addition, it has an advantage of fixing atmospheric nitrogen, providing a source of nitrogen for the following crop (Linemann and Azam-Ali, 1993). This is therefore a very good crop suited to peasant farmers for integration into their farming systems in the tropics where the ability of soils to retain inorganic fertilisers such as nitrogenous ones is low and such fertilisers are associated with high prices that most peasant farmers can not afford to buy due to lack of capital. Studies by Mukurumbira (1985), found that bambara groundnuts have higher residual nitrogen effect than groundnuts *(Arachis hypogea),* maize or fallow, consequently, there was no need to apply nitrogen to maize, for instance when grown after bambara groundnut in rotation. Again these findings offer a hint that the crop could be of value to small scale farmers in sub-Saharan Africa who are resource poor and cultivate crops in marginal, unproductive sites. The results for a trial performed by Mukumbira to ascertain residual effects of groundnuts, bambara groundnuts, maize, and fallow on the grain and stover yield of a subsequent maize crop is shown below:

Table 1. Residual effect of groundnut, bambara groundnut, maize and fallow on the grain yield and stover yield (kg/ha) of a subsequent maize crop.

	Groundnut	Bambara	Maize	Fallow groundnut
Grain	6190	7560	3880	4280
Stover	9740	12050	6690	7840

Adapted from Mukumbira (1985).

Bambara groundnut is also reported to be relatively free from pests and diseases and therefore its inclusion in the tropical cropping sequence (rotation) may delay serious build up of pests and diseases in the succeeding crop in the rotation. Thus it may reduce a need by resource poor farmers to buy pesticides to mitigate field crop pests whenever it is incorporated in peasant farming systems.

(b) Uses of bambara groundnut

Bambara groundnut is an indigenous grain legume grown mainly by subsistence women farmers in drier parts of sub-Saharan Africa. It is grown mainly for its seeds. Its seeds can be eaten fresh, when semi-ripe, as a pulse, when dry and mature or can be ground into flour (Linemann and Azam-Ali, 1993). The mature seeds are a rich source of protein ranging between 16% and 25% (Poulter and Caygill, 1980; FAO, 2001). In addition it contains the highest amount of lysine, one of the important amino acids, at a concentration of 400-430 mg/g of nitrogen (Key, 1979). It can therefore be a good substitute for expensive animal protein. Just like other legumes, it is deficient in methionine and cysteine. The lipid content of seeds is low (5-6% DM) (Poulter and Caygill, 1980; Aykroyd and Doughty, 1982). Total lipid content of seeds is inferior to other oil seed legume such as groundnut *(Arachis hypogea)* which has 45.3-47.7% DM (Rosario et al, 1981; Aykroyd and Doughty, 1982), but compare favourably with cowpea (1-1.6% DM), pigeon pea (1.2-1.5% DM) (Brough and Azam-All, 1992) and soy bean (Poulter, 1981). Brough and Azam-All, (1992) reported that proximate composition of bambara groundnut is not affected by adverse growing conditions such as water stress. From a nutritional point of view this property is of importance to the subsistence farmers who grow the crop in areas of low rainfall.

Bambara groundnut seeds can be used to produce vegetable milk that is comparable with soy milk and the ground seed can compete with, or replace, other conventional flours in a range of processed products (Poulter and Caygill, 1980). The tops of bambara groundnut are used as feed for livestock (Holm and Marloth, 1940; Duke et al, 1977). Furthermore, bambara groundnut has been used as protein source in pig and poultry ration (Uwaegbute, 1977).

Some key policy issues for consideration in relation to bambara groundnut

Reducing poverty is a primary focus of public policy in most sub-Saharan countries. One element of focus in formulating policy has been on improving welfare of rural community through improved agriculture. There are certain elements that governments of developing countries, especially sub-Saharan countries, should take into account when formulating policies that impact on livelihoods of rural peasants. These include a consideration of the proportion of the population impacted by proposed policy, the welfare implications of the policy particularly in regard to poverty reduction and economic growth, and the secondary effects of such policy, whether intended or unintended. It has been seen that bambara groundnut is cultivated mainly by women farmers living in rural areas. The policy should therefore target women farmers as potential beneficiaries. Moreover, from a nutritional point of view, bambara groundnut is seen to contain many essential nutrients as compared to other legumes. Governments of sub-Saharan Africa should formulate policies aimed at improving the crop so as to be able to tackle food security issues. There are possibilities to achieve sustainable development by formulating policies that are meant to promote under-utilised crops like bambara groundnuts. Sarris (2001) has given examples of several countries where evelopment in the nineteenth century occurred through a broad-based strategy to improve subsistence-oriented agriculture, a strategy that led to both reduced poverty as well as improved growth. The specific countries include: Japan, India, China, Bangladesh, Malaysia, Costa Rica and Indonesia. Sub-Saharan African countries can tape that example and promote such policies to alleviate poverty among its peasant farmers.

Conclusion

The aforementioned importance for bambara groundnut calls for a research effort to elevate the status of it from being under-utilised crop to a fully utilised crop. There is now convincing evidence that effective poverty reduction in subsistence farming can be met by promoting rare crops like bambara groundnut, which are grown by the disadvantaged women farmers. It is a hope that this paper opens a challenge to researchers to indulge deeply into assessing nutritional, processing and economic potential of bambara groundnuts in sub-Saharan Africa.

Acknowledgements

The preparation of this paper was based on a review of secondary materials, mainly research publications on bambara groundnut. I would like to specifically acknowledge a study by S. N. Azam-All; Karikari Sesay S. Karikari; F. J. Massawe; J. Aguilar-Mnjarrez; M. Bannayan; K. J. Hampson, and S. T. Collins whose paper titled 'Assessing the potential of an under-utilised crop - a case study using bambara groundnut' helped me a lot in drafting this paper.

References

Ameyaw, C. E. G and E. N. Duke. 1983. "The Effect of Soil Moisture Stress on Reproductive Efficiency and Yield of Bambara Groundnut (Voandzeia subterranean)." *Tropical Grain Legume Bulletin* 28: 23-29.

Aykroyd, W. R. and J. Doughty. 1982. "Legumes in Human Nutrition." FAO. *Food and Nutrition Paper* 20. FAO. Rome. Italy.

Brough, S. H. and S. N. Azam-All. 1992. "The Effect of Soil Moisture on the Proximate Composition of Bambara Groundnut (Vigna subterranean (L) Verdc)." *Journal of Science of Food and Agriculture* 60: 197-1163.

Doku, V.T. 1996. "Problems and Prospects for the Improvement of Bambara Groundnuts." Proceedings of the International Bambara Groundnut Symposium. 23-25 July 1996. University of Nottingham, UK.

Duke, J. A., B. N. Okigbo, C. F. Reed, and J. K. P. Weder. 1977. "The Bambara Groundnut (Voandzeia Subterranean Thours)." *Tropical Grain Legume Bulletin* 31: 4547.

Food and Agriculture Organisations. 2001. Agricultural Information Management Series.

Holm, J. M and B.W. Marloth. 1940. "The Bambara Groundnut or Njugu Bean." *Farming in South Africa* I5: 195-200.

Key, D. E. 1979. Crop and Production Digest No. 3. *Food Legumes*. Tropical Products Institute. London. pp 435.

Linnemann, A. R. and S. N. Azam-All. 1993. "Bambara Groundnut." *Pulses and Vegetables*, J. T. Williams, ed. Chapman and Hall. London. pp 13-58.

Masha, J. A. 2001. The Response of Bambara Groundnut (Vigna Subterranean (L) Verdc) to Moisture Regimes in Terms of Growth and Productivity. MSc. Theses. Sokoine University of Agriculture.

Mukurumbila, L. M. 1985. "Effect of Rates of Fertilizer Nitrogen and Previous Grain Legume Crop on Maize Yields." *Zimbabwe Agricultural Journal* 62: 177-179.

National Academy of Science. 1979. *Tropical Legumes: Resource for the Future*. Washington. pp 331.

Poulter, N. H. and J. C. Caygill. 1980. "Vegetable Milk Processing and Rehydration Characteristics of Bambara Groundnut (Voandzeia Subterranean (L) Thours)."Journal of Food and Agriculture 31: 1158-63.

Ramolemana, G. M.; G. S. Maphanyne, and M. Wassel. 1996. Response of Bambara Groundnut to Phosphorus with and without Irrigation." *Proceedings of the International Bambara Groundnut Symposium.* Nottingham University, UK. 23-25 July 1996. pp 231.

Rowland, J. R. J. 1993. *Dry Land Farming in Africa*. Chapman and Hall. London. pp 278.

Sarris, A. H. 2001. "The Role of Agriculture in Economic Development and Poverty Reduction" An Empirical and Conceptual Foundation Paper Prepared for the Rural Development Department of the World Bank Rural Action Plan 2001-2006.

Temu, H. E. M. 1994. Effect of Seed Size on Emergence, Growth and Development of two Landraces of Bambara Groundnut (Vigna Subterranean (L) Verdc). MSc. Theses. University of Nottingham.

Uwaegbute, H. O. 1977. "The Use of Local Feed Ingredients in Livestock Feeds." *Canadian Journal of Animal Science 5(7): 835.*

ASSESSMENT OF COPING STRATEGIES AMONG KENYAN WOMEN INALLEVIATING HOUSEHOLD FOOD INSECURITY

Olwambula Rose Ambalo

Introduction

Kenya is an agricultural country contributing to a quarter of the total gross domestic product (GDP). Over 50% of Kenya's population is engaged in subsistence farming of which 64% are women. Food is a complex multi-faced phenomenon with different properties arid meanings affecting everyone in several important ways, namely a household produce purchase or food received as gifts (FAO, 1998). According to the World Food Summit Plan of Action adopted in November 1996, food security exists when all people at all times have physical sufficient safe and nutritious food to meet their dietary needs and preferences for an active and healthy life (FAO, 1998). Sustainability of food supplies is the capacity to ensure the long-term stability of the household food supply and the ability of households to meet both the consumption and livelihood needs on a continuous basis (FAO, 1998). Household food security can only be ensured when the capacity to acquire food exists. A household acquires access to food through own production (Sharma, 1992). However, adequate availability of food at the national level does not necessarily translate to equitable distribution across the country or access among the households (FANTA, 2000). A household becomes food insecure when all the potential sources of food are strained or even threatened (UNICEF/ GOK, 1998). The indicators of household food security include budget surveys, food frequency assessments and food storage estimates. Globally 800 million people in developing countries do not meet their nutritional needs (FAO, 1998). While over 3 million people are dying of hunger in Kenya because the government has failed to attain food security and self sufficiency. Household food security has been deteriorating steadily with one in every two households in the Western, Coast, Eastern and North Eastern provinces classified as absolutely food poor (UNICEF/ GOK, 1998). According to Onian'go (2001), over 89% Kenyans are food poor, which means they are not adequately nourished. In Kenya, 48.7% of the populations live below the poverty line. The situation has even been exacerbated by the drought. The welfare monitoring survey data shows that about 50% of the rural and 29% of the urban population has been unable to access the minimum requirements of food and essential non-food commodities. These are the households classified by the FAO as absolute food poor (GOK, 1997; FANTA, 2000). In the urban areas, food is usually available but a nutritionally adequate diet is too costly for at least one third of the households (FANTA, 2000). According to the National Plan of Nutrition (GOK, 1994), many households experience both transitory and chronic food insecurity.

Predisposing factors to famine

Food security is basic to survival and sustenance of any nation. The government of Kenya recognises that meaningful development cannot be achieved without food security (GOK, 2000). Currently over 3 million Kenyans in different parts of the country are faced with starvation. The main food crops like millet, sorghum and maize did not produce good yield because of inadequate rainfall experienced in April and May in the affected areas of Rift Valley and Central province. However, regions like Budalang'i, Nvando and Suba West experience floods during long rain season and crops are washed away leading to food shortage. Rural agricultural mechanisation services like small irrigation projects have collapsed and water harvesting is not done in many parts of the country. Drought and floods have threatened peoples' lives, crops and livestock production. Kenyans heavily depend on agriculture, which is the backbone of our economy. The farmers reliance on rain for production results in a big blow on the economy and food security in times of drought or floods (Ochiel, 2004). Harvesting of water by communities to be used later for irrigation may be a solution to drought and floods as it will conserve the huge volumes that go to waste during the wet season. It will also provide a solution to weather like vagaries of drought which do cause conflicts when pastoralists move in search of water.

The famine situation in the country is so severe that the government has declared it a national disaster and has appealed for domestic and external food relief assistance to deal with the situation in the country. Interestingly, famine is not a new phenomenon in Kenya. The country has over the years faced increasing food deficits as a result of prolonged droughts and low agricultural productivity. This has exacerbated the food security situation in Kenya and increased reliance on food relief. It is a known fact that we are likely to experience a major drought every 10 years such as 1964, 1974, 1984, 1994 and 2004. We can almost be sure that unless we undertake sustained efforts, we will experience serious hunger around 2014 (Omiti, 2004). So, why do policy developers wait to be caught unaware and experience food shortages for four decades since independence? Worse still the government has been importing increasing volumes of foodstuffs like maize, wheat, rice and sugar into a country where agriculture contributes 80% of the employment and 60% of the national income without drought. This is an indication that policies, institutions and organisations have failed (Ochiel, 2004). The National Food Policy of 1994 does not empower communities in arid areas to seek alternative but environmentally friendly sources of livelihood. Famine is a national shame for scenes of starving Kenyans to be witnessed after independence. Food aid is not a sustainable venture in the country's economy (Omiti, 2004).

Food situation in the village of Kitise area in Makueni district was worsened with the outbreak of coccidiosis, a chicken disease that wiped out the population of the birds. The chicken trade has been a source of money to buy food for subsistence farmers to buy food. In the case of Turkana district, food shortage is not only attributed to erratic rains but also chronic insecurity that lead to frequent displacement of people. Women in Turkana have been growing crops along the river Turkwell in order to cope with the food situation in the area. However, they can no longer do so owing to the fluid and precarious security situation caused by incessant cattle raids executed by their Pokot neighbours. Thus hunger situation is aggravated by the insecurity along the Turkana and West Pokot boarder. Insecurity problem along the borderlines of the districts is a major factor in food insecurity and has contributed to persistent food scarcity leading to high poverty level in the district (Muga, 2004).

The unfavourable trade imbalance effected the farmers' retail negativeley while they had to purchase expensive fertilisers from abroad causing especially sufferings to small farmers. The statutory body in charge of maintaining strategic food reserves and relief supplies on behalf of the government, the National Cereals and Produce Board (NCPB), hardly attracts good maize intake from farmers due to the low prices they offer. There is stiff competition from private millers and traders who offer higher prices (Nyameino, 2004). Corruption is a significant factor in the agricultural setback. The NCPB reserves an average of 2 million bags of maize every year as part of the national food strategic reserves and famine relief food. Only when the maize harvest is exceedingly good and crop projections for the following season appear good, the government allows part of the strategic reserves to be exported. Corrupt parastal officers at NCBP exported part of Kenya's food reserves without following the normal tendering procedures, thus contributing to the current famine crisis. The Ministry of Agriculture approved the sell of half of 2 million bags. This was at the prevailing international market prices to offset part of NCPB's debt to farmers. Senior management officers of NCPB authorised the sell of the remaining maize without reference to the tender committee. The maize was sold at "negotiated prices" below those prevailing on the market at the time. The maize was not only sold at "uneconomical prices" which depleted Kenya's strategic food reserves that it led to the current famine (Inspectorate of State Corporations Report, 2004).

Food situation in Busia was made worse, when the Ugandan government froze exportation of cassava to Kenya. Though there was a bumper harvest in Kenya of about 29 million bags of maize last year, it coincided with famine in neighbouring Tanzania, and a significant quantity of maize was ferried to Central Tanzania which was hit by famine. Ordinarily huge amounts of maize entered Kenya as a result of cross-boarder trade with Uganda and Tanzania, where the cost of production is lower. However, last year there was a reversal of fortunes and Kenya supplied the grain to Tanzania.

Kenyans are also impoverished partly due to poor governance, which has contributed to the famine situation. How is the budget allocation of 17 billion shillings to the Department of Defence to be explained despite the fact that Kenya is not in a state of war, while the Ministry of Agriculture, which is the backbone of Kenya's economy, allocate only 5% to the budget. What is the point of protecting people from assumed external threat when they are faced with the real danger of death through hunger and malnutrition (Muga, 2004). It is interesting to note that as famine ravages different parts of the country, the poverty reduction strategy paper of 2001 ranks agriculture and rural development top an the list of priorities. Thus, the economic prospects for the year are particularly affected by famine and drought as this will push up food prices and drop the country into a high inflation regime (Omiti, 2004).

Abandonment of the guarantee minimum return loans in the 1970's and the low level of bank credit to the agricultural sector are disincentives to good food production. Kenya's vast agricultural land is potentially productive and is capable of not only feeding the nation but also a significant source of income to farmers. However, inadequate assistance arid the absence of enabling environment for women has stagnated the growth of agriculture. The majority of the 80% of rural population are women in fact more than half of the 14 million poor Kenyans in rural areas are women who experience low agricultural productivity, incomes limited employment opportunities, poor social and economic infrastructure.

Legal and cultural reasons limit women's right of access to credit (FAO, 1998). Men own land, which is usually required as collateral for a bank loan. Only 2% of arable land is owned or registered on the name of women. This limits the access to credit for the poor rural women. Making the women to be resource poor despite the role they play in food production, forestry and fishery. The access to or control over resources of production hardly tallies their increased responsibility, since they do not have access to land through ownership or secure tenure, which is a pre-requisite for improving agricultural production. Without secure land, the rights of the women have little or no access to credit or benefits of membership in rural organisations that are critical for obtaining inputs and services. Although women play a pivotal role in determining and guaranteeing food security and well being for the entire household, sadly, it is not counted as labour in statistics (Gitobu, 2004). Research has indicated that additional funds in the hands of women usually leads to better nutrition (FANTA, 2000). If the government ignores the role of women in food production, it ignores also the consequences of starvation (FAO,1998). Thus, equitable effective and sustainable agriculture and rural development cannot be pursued without an explicit recognition of the women's role in food security. There is need for promotion of gender based equity, access to arid control of productive resources (Gitobu, 2004).

Loss of soil fertility is one of the biggest constraints to maize growing in Western Kenya. Most soil types are acidic and contain low organic matter content rendering them poor for maize farming. The problem can be solved through multidisciplinary approach involving soil analysis, preventing soil erosion and addition of lime in organic manure to neutralise the acidity. Inappropriate technology has contributed to the country's problems of food production. Access to innovative technologies by the small-scale farmers will improve crop production (Nyameino, 2004).

Wild animals around Aberdare National Park and Maasai Mara Game Reserve are also leaving the game parks and damage crops in their search for water as a result of drought. They have contributed to poor harvest in the nearby districts this season. In Trans Mara District wild animals destroyed crops belonging to 500 farmers and they are petitioning the government to compensate them. Contrary, pastoral communities are moving all over looking for pastures, since watering points are drying up and livestock dying because of poor infrastructure before they reach the market for sale. This has led to acute food insecurity in several pastoral areas. The migration in search of pasture is a major source of conflict among the pastoralists' communities. Herdsmen searching for pasture and water have invaded game parks and private lands in Northern Kenya and the irrigation farms in Laikipia. This has caused human and wild life conflict.

Lack of land security like title deeds makes it difficult for women and the other small-scale farmers to engage in meaningful farming for food production. Availability of title deeds would act as security for bank loans and encourage investors in the area and this will create jobs for the locals. Other causes of food insecurity include household income, civil conflicts, poverty, diseases, land size, land tenure adequate labour, capital, rapid population and improved technology. Rural women's illiteracy and lack of information reduce their access to credit from financial institutions (FAO, 1998). At the moment food insecurity remains a national threat, because food availability has not kept pace with rapidly growing population (UOK, 1999). To cope with the drought many farmers in the famine hit areas resort to burning charcoal, hunting wild game and engaging in casual labour in their attempts to put food on the table. Some of these practices are counter productive as they increase the likelihood of drought and worsen its effects when it strikes.

The wide income distribution is a major setback to the agricultural growth. The limited Nations Development Programme (UNDP) in its 2004 Human Development Report paints a grim picture where 10% of the richest Kenyans control 36.1% of the country's wealth and 80% of the country's population share 48.8%. There is enough land and hard working people especially women in Kenya, but the lack of incentive is a cause for famine. Kenya is classified by the FAO as a low-income deficit country (FANTA, 2000). The food poverty line is below where people do not meet their minimum nutritional requirements. A major threat to the already inadequate food consumption of the poor is from the real income. Analyst have underlined that to deal with persistent hunger is to deal with poverty and underdevelopment (Kirti, 2000).

Studies done in Kenya by the International Policy Institute (IFPRI) showed that the absolute level of income is a key factor in the household food security. Other factors include the source, the control and the flow of income. As indicated earlier, women do not have access to adequate income. Many households are not able to produce enough food to carry them through to the next harvest and have to rely on purchased sources (World Bank, 1990). Women are responsible for food expenditure and spend the extra income on increasing food consumption (Brawn and Kennedy, 1986). It is hypothesised that if women have access to credit, they aare likely to invest in food production or in businesses or use it for purchase of food, making their households more food secure. Thus, financial sources are necessary for women if most households are to meet their food deficit.

Now that Kenya is facing another drought, it is expected that it has negative effects on the growth prospectives as famine triggers higher foods prices, escalating fuel and energy costs and imported inflationary on basic raw materials. Prices of livestock products have been falling sharply, sparking migration of herders as they search for greener pastures and water for their herd.

The role of women in household food security

It has been noted that most women head a large number of households as a result of absent husbands who migrate to towns in search of employment which in most cases is unavailable making women de facto heads of households (Mutahi & Onyango, 1994). FAO (1997) study indicates that in Africa income controlled by women is more likely to be spent on food. The same study revealed that at similar levels, households with women controlled incomes are more food secure than the male controlled incomes and that additional funds in the hands of women usually lead to better nutrition. Women are responsible for over 50% production of food grown world-wide and this is an important role they play in household food security. When they are not producing the food, they are earning the income for its purchase in order to enable their households to meet their daily nutrients intake (FAO, 1997; FAO, 1995). They are also involved in other crucial food security activities like preservation of biological diversity, processing and preparing food and care for the basic needs of the households (FAO, 1998). The indicators of household food security include budget surveys, food frequency assessments and food storage estimates.

It has been argued that women's control of income and other resources has significant influence on household food security than tht in the hands of men (FANTA, 2000). A study by Mutoro (1997) indicates that the main problem facing women is the lack of income due to the lack of credit. Women are not granted access to land, credit and other resources that would improve their productivity, which would enable them to generate income. Statistics show that women constitute 60% of the worlds' one billion poor living in absolute poverty and having no access to credit, thus having also caused the household's food insecurity in Kenya (Paul, 2000; UNICEF/GOK, 1998).

Coping strategies for food security

Women's income is becoming increasingly necessary to households of all types (Ongile, 1994). One strategy where women are fighting food insecurity is through the formation of women groups. These groups address social welfare issues and most of their activities are aimed at income generation. In Kenya, it has been noted that women are excellent entrepreneurs if they are given enough chances and space to start and operate small enterprises (Mutahi & Onyango, 1994). Thus, women groups in Kenya promote national development and provide different avenues for self-employment, which increase their purchasing power and access to emergency income. This increases family security and contributes to family welfare (UNICEF, I984). Various kinds of women groups exist almost in every community as formal and informal organisations in Kenya. These groups are attractive and useful to women of low-income households, since loans these groups get reach the poorest woman in the society. They provide income generating activities and investment opportunities for the poor and single women to manage their households in difficult situations (Mutoro, 1997). A study by Mutoro (1997) indicates that women groups in rural areas involved in income generation activities could improve their living standards and hence especially household food security, when the generated income is invested in agricultural production.

Another coping strategy is for households to produce their own food. Studies have shown that households that produce their own food are at a less risk of food insecurity than those that rely heavily on purchased food items (UNICEF/GOK, 1998). A household that does not produce food for consumption is subject to price risks arising from a sudden increase in the price of food it buys or a sudden fall in the price of the outputs it sells. A household can reduce the risk of being affected with price fluctuations by producing more of the food it consumes.

Income generation as coping endeavour increases food consumption and improves the quality of diet, hence improving household food security (Kennedy and Haddad, 1992). Women's enterprise activities should include credit, technology and training if they have to be effective.

Access to credit increases adoption to new technologies, such as improved seeds and chemical fertilisers and in turn it increases crop output, which is necessary to increase household income, thus reducing poverty and malnutrition (Babu, 2001). If the poor women were provided with additional income, more food would be demanded and produced. Where poverty is a cause of food insecurity and a country's government has enough income, it can purchase food for its citizens during periods of food shortages (Kirti, 2000).

Women have been a focus of development action projects concerned with women empowerment. Women organisations geared to income-generating activities can promote household development and provide avenues for self-employment (Mutahi and Mutoro, 1991). Thus, these organisations serve as successful coping strategies for women in rural areas by generating money and enabling them invest in agricultural production guarantee food security and well being for the entire household (Mutoro, 1997). If women's role in food production is ignored, the consequence is starvation.

When there is acute food shortage, women and children bear the brunt of famine as husbands leave starving families to look for jobs never to return (FANTA, 2000). Kenyans are entitled to decent living, dehumanising is to beg or watch children starve. With hunger biting, when food rations are handed out, women are forced to miss out on meals so that they can preserve food for children or wait for left over after the husband and children have eaten, they eat last. Women also provide for their families all types of palatable things some poisonous wild fruits like the wild cactus, which is now food for hundreds of impoverished families in Kilifi. People who cannot afford food eat wild fruits, dried and eaten like baobab and tamarinds (Kitui District Drought Management Officer, 2004). Yet, with the entire struggle women are disempowered by skewed development.

Families in Kitui, Makueni Machakos and Mbeere districts are faced with hard choice between hunger and possible death through the continued consumption of the contaminated maize in their store. Mid this year, over one hundred people in these areas died because there had been no alternative but to consume contaminated maize. By the time the government sent emergency relief food supplies to the affected areas over one hundred people had succumbed to the food poisoning. In severe incidences, women opt to scavenge for food for survival instead of watching their families suffer (Nyamaeno, 2004). In August it was reported that more than four people died in Kwale after eating contaminated fish, this had been scavenged from Kibarani dumping ground.

Another strategy for Kenyan women has been to look for odd jobs in the homes of the rich. In the event of unsuccessful job seeking mission women are forced take food on credit from the shop keepers and they pay back when the situation improves. Begging for food to feed their starving families is a coping strategy. Disposal of livestock through selling them cheaply is a common strategy, because pasture and livestock diminish fast at this time and it compels farmers to sell their animals at throw away prices. They argue that keeping the animals at home without pasture or water is demanding. Instead of struggling to provide for them most farmers are left with no option but to sell (Omita, 2004). Thus, the demand for grain pushes the prices up while the price of livestock of livestock decline. Other coping strategies for women in the affected areas is to produce charcoal and firewood to raise money for food. This is demoralising for a country that yearns for faster development as it affects the environment. Majority of those affected with famine are school going children. They are forced to stop attending school in search for food, although there is free primary education with a school feeding programme. As famine creeps across the land, most school children from poor rural families abandon school in search of work or food so that they can eat. In Ijara District, 13 public primary schools were closed indefinitely following the drought after pastoralists migrated with their children.

The government on its part has appealed for domestic and external food relief assistance to deal with current famine. It has also set aside one billion shillings for the purchase of relief food. This is not sustainable and for Kenya to break out of this cycle of begging, poverty and underdevelopment, sustained agricultural development is the key solution. The governments' rescue plan for this is to include long-term measures covering relevant sectors like agriculture, livestock water, sanitation, health and nutrition, which will cost 26 billion shillings (GOK, 2000). It will also cover those at risk of hunger. United Nations teams of experts in North Eastern and Coast provinces have sounded an emergency appeal to food aid to ball out the starving population in the provinces. Other humanitarian organisations are donating relief food in the affected areas to cope with the current food situation.

Conclusion

There is the suggestion that Kenya should adopt the biotechnology of genetically modified foods. This proposal is not suited for third world communities, where there are few other employment opportunities for those who lose their land and livelihood. People are hungry because they do not have enough to buy food. Recently in Kenya, farmers who had relied on forestland to cultivate crops were ordered to harvest their crops and leave. What will genetic crop seeds mean to these people? Maize and beans are reportedly the commonest staples in Kenya. On the other hand their production cost is high. The majority of rural Kenyans are unable to afford animal products or exotic fruits and vegetables. The current food insecurity in Kenya can be alleviated if people are encouraged to produce and consume variety of foods that are underexploited like the traditional foods such as sorghum, cassava, sweet potatoes and African green leafy vegetables. Indigenous vegetables constitute the cheapest source of the limiting macronutrients and it is apparent that traditional foods are of paramount importance in meeting the sustainable nutritional requirements for more than 80% of the rural Kenyan community.

Cultivation and popularisation of indigenous green leafy vegetables would be quite cheap and easy as the vegetables are already well accustomed to the harsh conditions of the tropics. These foods are crucial to food security particularly during famines and natural disasters. The plants grow as weeds in the wild or cultivated seeds and they need to be domesticated, since they require few inputs. Poorly informed agricultural policies have undermined women's ability to participate fully in food production activities. The development projects, government policies and support services appear to exclude women and this partly explains the impoverishment of female headed, female households in rural areas.

Girl child education is a long term intervention of household food security. The government should support women in their efforts on national development and eradication of poverty, by improving their education, health, and working conditions. Women should be given high priority in development spending and international aid policies if poverty eradication especially at household level is to be achieved in Kenya. The rural house-holds depend on tmely and adequate labour inputs so they need a source of cash income to be food secure through out the year. However, the economic development programs have not succeeded in helping women lift themselves and their family out of poverty. Micro-credit or small-scale employment can achieve poverty alleviation for women as this is increasingly seen as a method to improve general family welfare, health and nutrition. This will provide empowerment for an increasing number of women and contribute significantly to the amount of household income particularly in low-income households. Over reliance on the climate and foods bought at the market should be discouraged. Households should be assisted financially to enhance self-reliance in irrigation and food production.

Recommendations

A holistic approach and participatory methods need to be implemented:

- The government to create opportunities for co-operation between community-based (informal) and institution based (formal) systems for conservation and improvement of household food security. The government should improve its capacity to monitor and evaluate food supply situation with a view of reviewing the drought contingency plans and preparing to trigger response measure in affected sectors such as water, human and animal health.

- Policy makers should adapt agricultural policies and programmes that will lead to increased food consumption, diversification of diets, cultivation and consumption of nutritious food. This can be achieved through recognising and appreciating the central role of women. Thus a gender-based equity in access to and control of productive resources should be promoted.
- Security, law and order should be maintained in order to encourage investments in agriculture including agro processing and marketing is essential especially during peak harvest times of various food items to avoid high level of post harvest losses, because of pests, poor handling and unattractive market prices. Assist farmers to store their harvests properly to avoid their grains being contaminated.

References

Babu C. Suresh (2001) Food and Nutrition Policies in Africa: Capacity Challenges and Training Options. *African Journal of Food and Nutritional Sciences*.Vol.1. Washington.

Brawn V. and Kennedy, E. (1986). Commercialisation of Subsistence Agriculture: Income and Nutrition Effects. In: *Developing Countries* (Working Paper No.). Washington.

FANTA (2000). Potential Uses of Food Aid to Support HIV/AIDS Mitigation Activities in sub-Saharan Africa. Washington: Academy of Education Development.

FAO (1997). Agriculture Food and Nutrition for Africa. A Resource Book for Teachers of Africa. Rome: FAO.

FAO (1995) Assessing the Impact of Economic Policy on Food Security: A Training Manual. Rome: FAO.

FAO (1998). Rural Women and Food Securit Perspectives. Rome: FAO.

Gitobu, J. In Search of a Gender Sensitive Food Policy. Winrock International, Nairobi, Kenya.

GOK (1997). District Development Plan 1997-2001. Ministry of Planning and National Development. Nairobi: Government Press.

GOK (2000). Economic Survev. Ministry of Finance and Planning. Nairobi: Government Press.

GOK (1994). National Plan of Action For Nutrition. Nairobi: Government Press.

GOK (1999). Population and Housina Census.Vol. l. Nairobi: Central Bureau of Statistics.

Kennedy, E. and Haddad, P. (1992). Food Security and Nutrition 1971-1991. Washington.

Kirit, S. P. (2000). World Food System: Resilient for the Rich, Stubborn or the Starving: *Journal on Nutrition and Agriculture*. No. 20. Indira Gandhi Institute.

Muga, W. (2004) Lack of Policies is Kenya's Undoing. Nairobi, Kenya.

Mutahi, K. and Onyango, L. (1994). Support System from Women Employment in the Rural and Informal Sector in Kenya. Proceeding of the National Policy Workshop on Optimal Forms of Assistance to Women Employment in Rural and Infonmal Sectors. Nyeri, Kenya.

Nyameino, D. (2004). Cereals Growing Association of Kenya, Nairobi.

Ochiel, L. (2004). Women Critical to Food Security. Ace Communications. Nairobi, Kenya.

Ongile, G. (1994). Productive Employment and Equity in Rural and Informal Sectors in Kenya. Proceeding on the National Policy Workshop on Optimal Fonns of Assistance to Women Employment in Rural and Informal Sectors. Nyeri, Kenya.

Onian'go (2001). Enhancing People's Nutritional Status Throuh Revitalisation of Agriculture and Related Activities. *African Journal for Nutrition Science*. Vol.1. No.1. Pg.43-52.

Omiti, J. (2004). Seeking Long-term Solution to Food Shortage. Institute of Policy Analysis and Research. Nairobi, Kenya.

Pauli, L. (2000). Women's Empowerment and Economic Justice: Reflecting Experience in Latin America and Caribbean. New York. UNIFEM.

Sharma, R. P. (1992). Monitoring Access to Food and Household Food Security. Vol. 2. Rome: FAO.

UNICEF/GOK. (1998). Situation Analysis of Children and Women in Kenya. A Publication of GOK/UNICEF. Nairobi: Regal Press.

UNICEF (1984). Situation Analysis of Children and Women in Kenya: The Roles and Situation of Women. Central Bureau of Statistics: Nairobi.

PROSPECTS FOR ENHANCED FOOD SECURITY AND IMPROVED AGRICULTURAL SUSTANABILITY OF SMALL HOLDER SYSTEMS IN SUB-SAHAARAN AFRICA

Joshua O. Ogunwole

Abstract

Agricultural systems in sub-Saharan Africa (SSA) rest largely in the hands of the small holder farmers who practice under a highly variable rainfall condition and continuous soil nutrient depletion. Lack of financial resources has failed to make mineral fertiliser helpful to them. Alternative agriculture, an option to conventional method, is becoming more promising in terms of providing food security and safe environment. It involves the conversion of agricultural waste and manure into high quality biocides and soil amendments for plant use. In addition, an integrated natural resource- and plant nutrient management approach offer remedial response toward providing an enhancing strategy to food security and sustainability of small-holder system of agriculture in SSA.

Introduction

At the beginning of the new millennium, sub-Saharan Africa is confronting a number of daunting problems. These problems include extensive hunger, malnutrition, poverty, resource degradation and the spread of HIV/AIDS. As the name suggests, SSA refers to that part of the African continent, south of Sahara desert (i.e. south of latitude 20°N) or more precisely south of the isoline of 75 days length of growing period. The region is made up of 48 countries and covers 80% of Africa, i.e. 2960 million hectares of land with a population of over 500 million (INRA, 1995).

In SSA, the economic growth and quality of life depend largely on the agricultural sector, since majority of the population of the region remain dependent on agriculture for its livelihood. Agriculture accounts for more than 25% of the Gross Domestic Product (GDP) and it is the largest single employer of labour in the region (Abalu, 1990). Sub-Saharan Africa has been described as a region of small holder farmers. The agricultural system is predominantly of the small size of holding which ranges from 0.25 to 10 hectares (Ogungbile and Olukosi, 1990). It has often been referred to as a subsistence-, traditional- or a peasant-agriculture. This is because the primary motivation for cropping is family consumption and the excess, when available, for commerce.

Food production has declined or at best stagnated for the past 35 years. This period incidentally has been characterised by a steady increase in the population growth rate, which has averaged more than 3% per annum. Hence, the region remains the only area in the world where per capita food production has declined for more than two decades. Data from FAO (1997) showed that countries of SSA were unable to assure their citizens of a daily caloric intake of 2500 calories per capita resulting in deterioration in food security. According to the World Bank, about 30% of the region's population remains chronically undernourished (ECA, 1987). Dietary energy supply is believed to have fallen below the level required in about 30 countries in the region. In some of these countries, the nutritional deficit in terms of daily per capita food supply calories has been as high as 12% annually (ECA, 1987). This has exposed a growing number of people in the region to lethal diseases such as goiter and cretinism, with some of the victims dying in excruciating circumstances.

Given that the population of SSA is growing at about 3% per annum, the goal is to double agricultural output per annum to an average of 4 %. This is the minimum needed to permit modest annual improvement in per capita income. However, progress towards improved food security have been slow and uneven. Many countries in SSA have failed to make significant progress and some of them have actually suffered setbacks in their already fragile food security situation. The principal pressures threatening agricultural sustainability are those emanating from rural poverty, as more and more people attempt to extract a living out of dwindling resources. Efforts to support food security enhancement and agricultural sustainability in SSA will have to address issues relating to reduction in under-nutrition, combating rural poverty and achieving rates and patterns of agricultural growth that would contribute to overall economic development.

Poverty reduction and food security strategies must include measures to raise agricultural productivity, food production and distribution. These measures must be such that promote equal access for men and women to food, water, land, credit and technology.

According to Nwa (1997), the problem of food insecurity in SSA can be traced to four root causes of declining per capita food production, namely:

- The highly variable climate which has been observed to have a profound negative effect on the sustainability of agricultural production in the region, this undermines the pre-requisite for economic growth.
- Land degradation in the form of soil fertility depletion and soil erosion have adversely affected the intensification of land use systems in SSA.
- Biotic stress in form of weeds, pests and diseases which small holder farmers have to contend have been the cause of enormous losses for communities in SSA.

The lack of an enabling socio-economic environment such as limited or lack of market and market oriented production, inputs and implements, poor extension services, lack of access to credit and basic healthcare.

Climate and food production

One of the principal causes of fluctuation in food production in SSA is rainfall variability. Rainfall variability can be defined as the observed year-to-year differences in values of rainfall variables within an averaging period (i.e. about 30 years). Sub-Saharan Africa is characterised by a highly variable rainfall with coefficient of variability ranging between 15 to 30%. For example, at Samaru (Nigeria), the mean annual rainfall varied from 1481.8 mm in 1954 to 608.2 mm in 1983. Annual rainfall variability in SSA is over both time and space. A substantial proportion of the rains in SSA occur in a few high intensity storm ranging from 40 mm hr^{1} to as high as 120 mm hr^{1} in some instances. Variability also occur in the onset of the rains and the extent of early rainfall in SSA.

These variabilities have direct implications on the agricultural productivity and has a special urgency for the small holder systems of SSA, where per capita food availability has already declined consistently over the past three and half decades. The variability in the onset of the rains increases the risk in planting while delay onset will naturally delay planting and shorten the growing season resulting in fewer wet days (Jensen, 1990). For example, significant rainfall decrease have been reported over northern Botswana, Zimbabwe and eastern South Africa since the late seventies (Mason, 1996). Higher rainfall intensities throughout SSA accelerate soil erosion, which carry away vital nutrients from the topsoil and reduce the soils' organic matter (SOM) content. Global warming caused by the emission of greenhouse gases and increase in frequency of El-Nino events may have strong influence on the atmospheric circulation of the region with associated changes in rainfall patterns.

Agriculture is usually well adjusted to mean climatic conditions of a region. However, this ability to adjust declines as conditions become progressively extreme and crops become more vulnerable to climate variability (Sivakumar, 1997). Persistence of the current trends in SSA would mean drier areas and more deserts in the long run. The most common consequence of rainfall variability is drought, which increases crop failure resulting in hunger, poverty and famine. Water conservation is one adaptive strategy to cope with rainfall variability and to reduce its influence on food production.

In SSA, high evaporation rates are common and this coupled with the unpredictable rainfall often lead to periods of water shortage in some areas particularly, the semi-arid. When rainfall exceeds potential evapotranspiration, soil moisture reserves are recharged and more water is available for plant use. However, when rainfall is less than potential evapotranspiration, soil moisture reserves are utilised and stress symptoms manifest in plants which significantly alter their productivity. Strategies that must be adopted must be those that will mitigate accumulation of greenhouse gases and those that will adapt to the consequences of climate variability and change.

Soil fertility degradation

A major development concern in SSA is the feeding of its rapidly growing population. Population is increasing in a region where, paradoxically, the soils are fragile with low inherent fertility. Most soils of the region are highly weathered, characterised by a low activity clay with a low cation exchange capacity. In addition to these is their high susceptibility to crusting, compaction (Perrier, 1987), runoff, erosion and the leaching of plant nutrients (El-Swaify et al, 1987). Soil fertility degradation has been described as the single most important biophysical constraint to food security in SSA (Gicheru et al, 2003).

The traditional method of maintaining soil fertility is to allow land to revert to natural vegetation after two or three years of continuous cropping (referred to as bush fallowing). This restoration practice is stable and biologically efficient. However demographic and economic changes have caused a widespread disappearance of this soil fertility restoration practice and cultivation has expanded into marginal soil types.

The result of all these is systematic degradation of many land holdings and declining yield of crops. During the last three decades, depletion has been estimated at an average of 660 kg/N/ha, 76 kg P/ha and 450 kg K/ha from about 200 million hectares of cultivated land in 37 countries in SSA (Sanchez, as quoted by Spore, 2003). Too many nutrients are fast being lost in these soils resulting in a negative nutrient balance for many dropping systems. There is a need to arrest this situation.

One obvious solution to solving soil fertility problem in SSA is to import nutrients in form of mineral fertilisers. However, for a variety of socio-economic and political reasons this is generally difficult; the resource poor farmers in this region do not have the financial resources to take advantage of mineral fertilisers even when they are available. Report has it that average mineral fertiliser use in SSA is 9 kg/ha compared with a global average of 98 kg/ha (Spores, 2003).

Biotic stress

The intensification of land use systems in SSA is facing increased biotic pressure. Farmers' fields are getting heavily infested with pests and diseases after only few years of cropping. Agricultural pests include weeds, insects, rodents, nematodes and plant pathogen. One very important cause of stress to cultivated crops, which has adversely affected the livelihood of some 100 million people in SSA is the parasitic weeds of the genus Striga (Adetimirin and Kim, 1999). Striga hermonthica and S. asiatica parasitize staple crops like maize, sorghum, rice and millet. Striga gesneroides parasitizes cowpea. The damage caused by striga in SSA is greatest in the savannah areas which extend from Cape Verde on the west coast through central, eastern and southern Africa. An estimate of 21 million hectares of farmland devoted to cereal production are already infected by Striga (Kroschel et al, 1999). In the Northern Guinea savannah agroecology of Nigeria estimated yield loss is as high as 46% with an average of 10% from 66 fields (Oikeh et al, 1996). Studies from six West African countries reported yield loss of 24% from all striga spp (Sauerborn, 1991). Losses of up to 100% occur on some farmers' field. The shorten fallow periods due to demographic growth had resulted in hosts of Striga spp. being cropped more frequently favouring their reproduction and spread. Striga menace is fast spreading among farmers' fields and could spread even to new areas if control measures are not taken in time.

Socio-economic constraints

The lack of adequate, affordable and reliable infrastructural services like good roads, water supply, sanitation and energy in SSA touches the lives of the rural families daily. Women and children would have to walk very long distances to the nearest source of water which may be unsafe for drinking. Rural women cook with locally manufactured stoves which are inefficient causing their kitchens to be filled with smokes that causes severe damages to their lungs and eyes.

Most of the labour needed for agriculture in SSA is provided by women. This is because most men migrate to cities in search of Jobs or are conscripted into the army in conflict prone countries. The women have the primary responsibility for food production. Yet, they do not have access to resources that would enable them increase output. Lack of means of transport, good roads, access to markets, inputs and vital information coupled with lack of electricity are recurrent problems of SSA countries.

Civil and international conflicts present enormous challenges to sustainable agriculture and food security in SSA. Conflict takes land out of production through direct displacement of farmers or placement of mines. Resources are channelled to the armies to prosecute campaigns. Millions have been killed or wounded or rendered homeless and deprived of their livelihoods. In Eastern and Southern Africa alone, conflicts have pushed millions of the people to the verge of starvation with women and children being the most hit. Conflict has destroyed incentives to invest in farming business in most of the conflict-prone-countries in SSA.

Malaria and HIV/AIDS are the leading killer diseases in SSA, whose effects on agriculture and rural life are too sadly evident in daily lives (Spore, 2002a). Many agricultural practices have the effect of encouraging breeding grounds for malaria disease vectors. Large-scale dams built for water and energy along with the small earth dams and fish ponds currently favoured for irrigation and protein are habitat to the insect larvae. Malaria has an adverse effect on agricultural production. It has reduced Gross National Product (GNP) of countries in SSA by more than 1% (Spore, 2002a), causing spells of illness and premature death. HIV/AIDS has orphaned more than 10 million children in SSA (Spore, 2002b). AIDS undermines agriculture and affects food security of rural families. Families spend more on medical bills and funeral expenses thereby reducing the little resources available for farming. Women and girls are the most vulnerable to HIV/AIDS and face the greatest burden of caring for the sick and orphaned children. Any support to the region must be such that would assist it in greater utilisation of its agricultural base in order to increase overall production.

Strategies for fostering agricultural productivity among small holder farmers of sub-Saharan Africa

Agricultural production in SSA is primarily conducted by small farmers, who are very conservation conscious but have limited resources (Twomlow et al, 1994). Early efforts to support and advance agricultural development in SSA were heavily based on transferring technologies and management practices of the developed countries. The age-long traditional practices of the farmers which has a high resilience to weather fluctuations, pest attacks and soil degradation were commonly discouraged and replaced by the transferred technologies which were less stable. It is now becoming widely accepted that the shift was undesirable in many situation and deleterious in others for both short-term food security and long-term sustainability in most of the environments.

To 'leap-frog' agriculture in this region, there is a need to develop a pro-active and coherent response to the challenges posed by climate variability, soil fertility degradation, biotic stress, etc. and to refocus agricultural research capacity within the countries in SSA. The approach to fostering agricultural productivity encompasses two sets of strategies: those to mitigate the effect of climate (rainfall) variability and soil fertility depletion and those to adapt to the consequences of these constraints. Among the mitigation strategies are increasing carbon stock in the agro ecosystems, promoting increased carbon sequestration through improved management of forests, crop- and grazing-lands (Metting *et al,* 1999), improving nitrogen use efficiency and reducing nitrous oxide emissions. The adaptive strategies include pursuing integrated resource management for higher yields and enhancing germplasm for better tolerance of abiotic and biotic stresses. Others are developing approaches to cope with greater climatic variability and increased frequency of extreme weather events. These strategies must be environmentally friendly, readily available, affordable and sustainable for easy adoption by the farmers of the region.

Conservation tillage

The use of tractor to plough and harrow soil to a depth of 0.2 to 0.3 m every season has spread to much of the region in the last two decades. The need to break-up cultivated layer, increase soil and water conservation (by improving infiltration) and weed control (deep burial of weed seeds) are some reasons adduced for deep ploughing. This cultivation practice is one of the principal causes of soil fertility degradation. The practice causes loss of soil organic matter (SOM) and increase loss of soil carbon in the form of CO_Z (a greenhouse gas), which consequently contribute to yield reduction.

An alternative to this cultivation practice across ecologies in SSA is to introduce conservation tillage (i.e. minimum and no-tillage system). Conservation tillage involves the drilling of seeds directly into the ground. It may involve shallow cultivation using tyned implement which disrupt the soil to a depth of only 0.1 m. This tillage practice minimise soil disturbance and ensure presence of a residue cover which is an important contribution to SOM in the arid and semi-arid areas of SSA. No-till reduces soil erosion and carbon losses (relative to conventional tillage) by reducing carbon mineralisation rates. This is possible because the soil is less disturbed and experiences little or no exposure to radiation. With careful soil management, as reported in literature, the potential sequestration of carbon per year may reach 0.05 - 0.1 $Mgha^{-1}$ for rainfed crop- and range-land (Rosenberg et al, 1999). This tillage system has the potential to mitigate the greenhouse effect. In addition, high SOM levels lead to improved soil structure, increased water ifiltration and increased soil water holding capacity and water use efficiency of cropped plants.

Water harvesting techniques

Given the erratic and variable rainfall particularly, in the arid and semi-arid areas of SSA, and the limited possibilities of increasing cultivated areas, the priority of agriculture across all dry land farming systems of SSA is to increase biological and economic yield per unit of water (Van Duivenbooden et al., 1999). One proven strategy to adapt to the variable rainfall scenario is to harvest rain water in the root zone of crops and manage field and crops through mulching, to use water more efficiently.

Water harvesting is the collection and concentration of rain water runoff for increase and more reliable agricultural production. The principle here is harvesting and utilisation of runoff instead of allowing it to cause erosion. One popular water harvesting technique is the tied-ridge system. The technique has been reported to be a reliable and sustainable method of increasing yield of maize (Hullugalle, 1989) and cotton (Ogunwole, 2004) in the West African savannah. In Zimbabwe, sheet erosion was negligible when tied ridge techniques was performed on semi-permanent ridges (Vogels, 1993). Other techniques include traditional terraces, planting pits or 'Zai' and stone bunds (Reij et al, 1996).

Soil fertility restoration

The search for options to arrest soil fertility degradation in SSA is being intensified. Any acceptable strategy must be such that will increase agricultural production and safeguard the environment. The application of mineral fertilisers is an effective technique, for increasing crop yields in arable farming in SSA (Gigou and Bredoumy, 2003). However, the use of these fertilisers alone will lead to decreasing crop yields in the long run (Greenland, 1994). Maintaining SOM is a key to sustainable land use management in SSA, because SOM acts as a source and sink for plant nutrients (Bationo *et* al, 2003). Therefore, soil management strategies that will improve and maintain SOM will increase and sustain crop yield.

Alley cropping of food crops with shrubs or trees conserve soil, recycle nutrient into the soil and supply food and fuel to small farmers. This is a proven soil management strategy particularly for humid regions of SSA. This system can replace the long fallow or rotation practices and ensure overall stability in food crop production. In West Africa, *Leucaenia leucocephala* and *Gliricidia sepium* are gaining widespread acceptance on non-acid Alfisols, while *Sesbania sesban* is the favourite among east African farmers (Mafongoya et al, 2003). These shrubs are planted in hedgerows and pruned about five times a year to provide mulch and green manure for companion food crops and at the same time prevent shading of the food crops. Agroforestry, which is the planting of trees on farms and in the entire rural landscape, can increase the resilience of farming systems and sequester significant amount of carbon - across all ecologies of SSA. Land clearing and continuous cultivation in most small holder systems of SSA have decreased soil carbon storage and fertility (Swift et al, 1994). Many agroforestry systems using fast growing, nitrogen-fixing trees can reverse this trend. Agroforestry systems restore degraded soils through recovery of organic based nutrient cycles by replenishing SOM.

In West and Central Africa and some parts of East Africa, the natural recycling of animal dropping through its disposal on cultivated land for crop production is a good soil fertility management strategy from the standpoint of improving soil's productive capacity and reducing the dependence on mineral fertilisers (Ogunwole and Ogunleye, 2004). Animal manure increases SOM, nutrient availability, cation exchange capacity and water holding capacity of soils. It also increases yields of crops when sufficient quantity is applied on a continuous basis. Animal manure application to crop land are of several forms. Farmers may gather manure from stalls, transport and hand spread it on field or corral their animals overnight on field during the dry season. The latter form returns both manure and urine to soils and results in greater crop yields than the former (Bationo et al, 1997). Corralling also require no labour for manure handling, storage and spreading (Bationo et al, 1997). The superiority of animal manure has been attributed to their slow release of balanced nutrient reserves during decomposition, effects on physical properties, increasing the supply of trace elements and amelioration of the effects of acidifying fertilisers under continuous cultivation.

The biological nitrogen fixation approach to soil fertility improvement particularly, in the West African savannah with the development of sustainable maize legumes rotation is another proven strategy for soil fertility maintenance in SSA (Anthofer and Kroschel, 2002). Well adapted varieties of soybeans are used effectively for SOM and nutrient replenishment through production of large amount of biomass and fixing higher amount of nitrogen yet with a tolerance to low phosphorus. Reports have it that this strategy has not only resulted in a positive nitrogen balance in maize rotations and significantly increased maize yield (Carsky and Berner, 1997), but has increased gross income of farmers by about 30 to 70% compared to farmers that continued on solecropped maize in Northern Guinea Savannah agroecology. Integrated Nutrient Management approach with sound combinations of mineral fertiliser and organic inputs is a central strategy to soil fertility management and restoration in SSA (Vanlauwe *et al,* 2002).

Integrated pest management

The use of pesticides by small holder farmers in SSA is limited, due to their relatively high costs and their non-availability in most local markets. In commercial food crops and industrial crops where chemical control has been used abundantly, often result in increased short term yields, human poisoning, development of pest resistance to pesticides and water pollution.

An alternative approach to pesticide use is the holistic approach to pests and disease control which integrates a wide range of cultural practices. Such practices are: (i) crop diversity through rotation and mixed cropping; ii) organic manuring to stimulate soil biological activity and (iii) careful use of selected biological control techniques, naturally occurring plant extracts and materials. This approach is known as Integrated Pest Management (IPM) since it tackles the entire agro-ecosystem in which both the crop (or livestock host) and pests are but two components. Different tactics are used in developing site-specific management system that capitalise on natural biological factors that serve to maintain the pests at levels below those causing economic loss. Report from Amhara region of Ethiopia has it that fermented cow's urine and sisal juice significantly reduced the prevalence of endemic bush crickets, while wild onions juice kept rodents out of crops and crocodile climber plants kill weevils that stored grain in an organic integrated pest management project (Spore, 2002b). In the dry savannah of West Africa, extracts of neem *(Azadirachta indica),* garlic *(Alium satirum L.)* and black eye pepper *(Piper guinensis)* have been reported to have biocidal potentials (Alabi & Olorunju, 2004). In Kenya, the combination of *Desmodium,* a fodder legume, in an intercrop with striga tolerant maize and Napier grass is fast checking incidence of stem-borers and striga. Napier grass traps stem-borer moths while *Desmodium* suppresses striga and improve soil fertility. Over 20% increase in maize yield has been reported as a result of the system. In Nigeria, intercropping striga suppressing varieties of soybean, cowpea and groundnut with striga tolerant cereals has increased striga suppression by about 60% (Carsky and Berner, 1997).

Support systems for fostering agricultural productivity

Food security and agricultural sustainability can be achieved if the leaders of the countries in SSA invest in infrastructure that will improve the linkages between rural areas and small town, and trading centres. They must also enhance productivity and raise the quality of life of the rural population through provision for education, healthcare, clean water and sanitation system. A good network of transport, power and telecommunication services must also be provided. Although, African leaders have increasingly expressed renewed commitment and determination, through the New Partnership for Africa's Development (NEPAD), for agriculture and rural development. There is a need for them to pursue more intensive efforts in the years ahead. Such efforts can lead to a rapid growth that has benefited rural people in other developing countries of the world notably India and China. In addition, there is a need to empower women in agriculture. Governments and Non-governmental Organisations (NGOs) could help organise women into groups to gain access to resources. Such groups may further be assisted to provide clean water, construct toilets and improve family health. The group can also be facilitated to play dominant role in seed production and agricultural fairs. Results of an analysis in Zambia showed that if women commanded the same factors of production as men, output in Zambia would increase by up to 15% (IBRD, 2002).

Conclusion

In SSA, where over 30% of the population remains chronically undernourished, reliance on agriculture holds promise for poverty alleviation and food security. Consequently, achieving a sustainable food security in the region requires strengthening the agricultural sector and reversing the degradation of the natural resources upon which agriculture depends. Strategies to sustainably intensify agriculture in less favoured are preferred. There are also areas that are inadequately linked to infrastructure and services such as market and health centres.

Water harvesting techniques (like the tied ridge practices) offer potential to secure agricultural production and reduce financial risk associated with the use of mineral fertilisers. Under the conditions of adequate water supply, the addition of mineral fertilisers and organic amendments will increase crop water use efficiency. Integrated nutrient management is an efficient and judicious way to increase agricultural production and sustain land use. The practice of alley cropping and agroforestry are also of importance. In addition to soil fertility restoration and improvement, agroforestry enhance more carbon sequestration and increase biomass production for livestock and soil mulching. However, given the complex and interlinked components of the challenge to feeding SSA, it is clear that strategies that deal only with one part (i.e. crop productivity or land use or water conservation for instance), will not be sufficient. The issues are interrelated and must be dealt with as an interlocking holistic system.

References

Abalu, G. O. I. 1990. "The Attainment of Food Security in Nigeria: The Role of Resource Constrained Nigerian Farmers." J. O. Olukosi, A. O. Ogungbile, and B. A. Kalu, eds. *Appropriate Agricultural Technologies for Resource-Poor Farmers*. pp. 520.

Adetimirin, V. O. and S. K. Kim. 1999. "Farm Level Options for Controlling Striga on Maize in Africa." In: J. Kroschel, H. Mercer-Quarshie, and J. Sauerborn, eds. *Advances in Parasitic Weed Control at On-farm Level.* Vol. 1. *Joint Action to Control Striga in Africa.* Margraf Verlag, Weikersheim, Germany, 89-101.

Alabi, O. and P. E. Olorunju. 2004. "Evaluation of Neem Seed Extract, Black Soap and Cowdung for the Control of Groundnut Leaf Spot at

Samaru, Nigeria." *Archives of Phytopathology and Plant Protection* (37): 123-127.

Anthofer, J. and J. Kroschel. 2002. Partial Macronutrient Balances of Mucuna/maize Rotations in the Forest Savannah Transitional Zone of Ghana. In: B. Vanlauwe, J. Diels, N. Sanginga, and R. Merckx, eds. *Integrated Plant Nutrient Management in Sub-Saharan Africa.* 87-111.

Bationo, A., T. O. William, and A. U. Mokwunye. 1997. "Soil Fertility Management for Sustainable Agricultural Production in Semi-arid West Africa." In: T. Bezuneh, A. M. Emechebe, J. Sedgo, and M. Ouedraogo, eds. *Technology Options for Sustainable Agriculture in Sub-Saharan Africa.* 349-367.

Bationo, A., U. Mokwunye, P. L .G. Vlek, S. Koala, and B. I. Shapiro. 2003. "Soil Fertility Management for Sustainable Landuse in the West African Sudano-Sahelian Zone." In: M. P. Gichuru et al, eds. *Soil Fertility Management in Africa: A Regional Perspective*, 253-292.

Carsky, R. J. and D. K. Berner. 1997. "Benefits of Crop Rotation with Soybean and Cowpea in Savannah Cereal Based Systems." In: T. Bezuneh, A. M. Emechebe, J. Sedgo, and M. Ouedraogo, eds. "Technology Options for Sustainable Agriculture in Sub-Saharan Africa," 391-402.

ECA (United Nations Economic Commission for Africa). 1987. Socio-economic Constraints to Food Production and Utilisation. Regional Perspectives. Paper Presented at the Regional Seminar on the Role of Chemistry in Improving Food Supplies in Africa. 9-12 March. Nairobi, Kenya. ECA, Addis-Ababa, Ethiopia.

El-Swaifv, S. A., S. Singh, and P. Pathak. 1987. "Physical and Conservation Constraints and Management Components for SAT Alfisol." In: P. Pathak, S. A. El-Swaify, and S. Singh, eds. Alfisols in the Semi-Arid Tropics. Proceedings of the Consultants' Workshop on the State of Arts and Management Alternative for Optimizing the Production of SAT Alfisols and Related Soil. 1-3 Dec. India.

FAO. 1997. *Report of the World Food Summit, 13-17.* Part One. November, 1996. Rome, Italy. Food and Agricultural Organization of the United Nations.

Gicheru, M. P., A. Bationo, M. A. Bekunda, H. C. Goma, P. L. Mafongonya, D. N. Mugendi, H. M. Murwira, S. M. Nandwa, P. Nyathi, and M. J. Swift. 2003. *Soil Fertility Management in Africa:*

A Regional Perspective. Academy Science Publishers, Nairobi, Kenya. 306p.

Gigou, J. and S. K. Bredoumy. 2002. "Yield Trends, Soil Nitrogen and Organic Matter Content during 20 Years of Continuous Maize Cultivation." In: B. Vanlauwe, J. Diel, N. Sanginga, and R. Merckx. eds. *Integrated Plant Nutrient Management in Sub-Saharan Africa.* 199-208.

Greenland, D. J. 1994. "Long Term Cropping Experiments in Developing Countries: the Need, the History and the Future." In: Leigh, R. A. and A. E. Johnston, eds. *Long Term Experiments in Agricultural and Ecological Sciences.* CAB International, Wallingford, UK, pp 187-210.

Hullugalle, N. R. 1989. "Effect of Tied Ridges and Undersown Stylosanthes Hamata L on Soil Properties and Growth of Maize in the Sudan Savannah of Burkina Faso." *Agriculture, Ecosystems and Environment* 25: 39-51.

Institute for Natural Resources in Africa (INRA). 1995. Land Restoration and Integrated Watershed Management in Sub-Saharan Africa: A Project Proposal. Institute for Natural Resources in Africa of the United Nations University, Accra, Ghana.

International Bank for Rural Development (IBRD). 2002. From Action to Impact: the Africa Region's Rural Strategy Rural Development Operations, the African Region 49pp.

Jensen, J. R. 1990. Decreases in Point Annual Rainfall in Northern Nigeria, 1905-1983. (89). Paper Presented a the First Biennial Hydrology Symposium, "Hydrology for National Development", IHD Programme of Nigeria. Maiduguri, Nigeria.

Kroschel, J., H. Mercer-Quarshie, and J. Sauerborn. 1999. "Preface, Advances in Parasitic Weed Control at On-Farm Level." *Joint Action to Control Striga in Africa.* Margraf Verlag.

Mafongoya, P. L., D. N. Mugendi, B. Jama, and B. S. Waswa. 2003. "Maize Based Cropping Systems in the Subhumid Zone of East and Southern Africa." In: M. P. Gichuru et al, eds. *Soil Fertility Management in Africa: A Regional Perspective.*

Mason, S. J. 1996. Regional Manifestation of Climate Variability in South Africa. Proceeding of Workshop on Reducing Climate-related Vulnerability in Southern Africa. October 1-4 Victoria Falls, Zimbabwe. 19-25.

Metting, F.B., J. L. Smith, and J. S. Amthor. 1999. "Science Needs and New Technology for Soil Carbon Sequestration." In: N. J. Rosenberg, R. C. Izaurralde, and E. L. Malone, eds. *Carbon Sequestration in Soils: Science, Monitoring and Beyond.* Proceedings of the St. Michaels Workshop.

Nwa, E. U. 1997. Food Security Issues in Sub-Saharan Africa. A Keynote Paper Presented at the Joint START/WCRP/SCOWAR Workshop on "Climate Variability Prediction, Water Resources and Agricultural Productivity: Food Security Issues in Tropical Sub-Saharan Africa". Cotonou, Benin, 22-25 July.

Ogungbile, A. O. and J. O. Olukosi. 1990. "An Overview of the Problems in the Resourcepoor Farmers in Nigerian Agriculture." In: J.O. Olukosi, A. O. Ogungbile, and B. A. Kalu, eds. *Appropriate Agricultural Technologies for Resource-Poor Farmers*. 21-34.

Ogunwole, J. O. 2004. "Effects of Fertiliser and Time of Ridge-tie on Yield and Fibre Quality of Late Sown Cotton in the Dry Savannah Zone of Nigeria." *Journal of Sustainable Agric*. 24(3): 97-107.

—— and P. O. Ogunleye. 2004. "Surface Soil Aggregation, Trace and Heavy Metal Enrichment under Long-term Application of Farmyard Manure and Mineral Fertilisers. *Communications in Soil Science and Plant Analysis* 35(11 & 12): 15051516.

Oikeh, S. O., G. Weber, S. T. O. Lagoke, and A. E. Awead. 1996. "Assessment of Yield Losses from Stria Hermonthica in 'Farmers' Fields in the Northern Guinea Savannah." *Nigerian Journal of Weed Science* (9) 1-6.

Perrier, E. R. 1987. "An Evaluation of Soil-water Management on an Alfisol in a Semi and Tropics of Burkina Faso." In: P. Pathak, S. A. El-Swaify, and S. Singh, eds. Alfisols in the Semi-arid Tropics. Proceedings of the Consultants' Workshop on the State of the Art and Management Alternatives for Optimizing the Productivity of SAT Alfisols and Related Soil, 1-3 December, 1983. ICRISAT Center, India. Patancheru, AP 502 324, India.

Reij, C., I. Scoones and C. Toulmin. 1996. *Sustaining the Soil: Indigenous Soil and Water Conservation in Africa.* Earthscan Publications Ltd. London. 260pp.

Rosenberg, N. R., R. C. Izaurralde and E. L. Malone. 1999. Introduction to Proceeding on Carbon Sequestration in Soils: Science, Monitoring and Beyond. Proceedings of the St. Michaels Workshop pp. vü-xü.

Sauerborn, J. 1991. The Economic Importance of the Phytoparasites Orobanche and Striga. In: J. K. Ransom, L. J. Musselman, A. D. Worsham, and C. Parker. eds. Proceedings of the 5th International Symposium of Parasite Weeds, 137-143.

Sivakumar, M. V. K. 1997. Climate Variability and Food Production. Scientific Lecture Delivered at the 49th Session of the Executive Council of the World Meteorological Organisation. 10-20 June. Geneva, Switzerland.

Spore (2002a). Poor Health and Poor Agriculture. Number 97: 1-2. A CTA Publication.

—— 2002b. What's that Smell? Number 101: 6. A CTA Publication.

—— 2003. Women in Charge Number 103: 6. A CTA Publication.

Swift, M. J., P. D. Seward, P. G. H. Frost, J. N. Qureshi, and F. N. Muchena. 1994. Long Term Experiments in Africa: Developing a Database for Sustainable Land Use Under Global Change. In: R. A. Leigh and A. E. Johnston, eds. Long-term Experiments in Agricultural and Ecological Sciences. CAB International, Wallingford, U.K: 229-251.

Twomlow, S., T. Mashavira, T. Willcocks, H. Dhliwayo, W. Gapaza, and P. Hynes. 1994. Conservation Tillage Practices for Resource Poor Farmers in Zimbabwe. Presented at 2nd International Symposium on Sealing, Crusting and Hardsetting Soils: Productivity and Conservation. University of Queensland. Brisbane, Australia. 7-11 February.

Van Duivenbooden, N., M. Pala and M. Doumbia. 1999. An Overview of the Optimising Soil Water Use Consortium and Objectives of the Workshops. In: N. Van Duivenbooden, M. Pala, C. Studer, and C. L. Bielders, eds. Efficient Soil Water Use: the Key to Sustainable Crop Production in the Dry Areas of West Asia, and North and Sub-Saharan Africa. Proceeding of Workshop on Optimising Soil Water Use Consortium. Aleppo, Syria: ICARDA. Patancheru, India: ICRISAT. pp. 1-14.

Vanlauwe, B., J. Diels, N. Sanginga, and R. Merckx. 2002. Introduction; In: B. Vanlauwe et al, eds. *Integrated Plant Nutrient Management in Sub-Saharan Africa*: 1-5.

Vogels, H. 1993. "Tillage Effects on Maize Yield, Rooting Depth and Soil Water Content on Sandy Soils in Zimbabwe." *Field Crops Res.* 33: 367-384.

STRATEGIES FOR THE DEVELOPMENT OF SMALLHOLDER AGRICULTURE AND HEALTH IMPROVEMENT IN SUB-SAHARAN AFRICA

Adesola Ajayi

Introduction

Agriculture plays a significant, irreplaceable role in the economies of all nations in the world. It was the backbone of every civilisation that human societies have passed through beginning with human transition from nomadic gatherers to sedentary growers of plants. Modern society is absolutely dependent on agriculture for food and the society places a great emphasis on the availability of sufficient quantities and qualities of food, thereby necessitating the current global efforts to combat food insecurity, hunger and malnutrition. The importance of agriculture in sub-Saharan Africa lies not only in its contribution to the GDP but also in the survival of the majority of each country's population that live in the rural areas and whose livelihood is totally dependent on agriculture. These people practice agriculture under extremely harsh climatic/environmental conditions with limited access to basic production-enhancing inputs, incentives and competitive markets. In the light of current trends of globalisation and mergers, these poor but enterprising people are systematically rendered more vulnerable in what has now become a global village. There is therefore an urgent need for practical, sustainable and relevant strategies for improving agriculture and the livelihoods of the millions of poor people whose lives and future are totally dependent on it.

Agriculture in sub-Saharan Africa today

More than 90% of farmers in sub-Saharan Africa, especially in the western and eastern parts, are smallholders. Collectively, this group constitutes a larger fraction of both the total agricultural workforce and the output. They use traditional and manual methods to grow crops, often in complex mixtures and in well-proven farming systems (Barret et al, 2002). They produce nearly all the food requirements of each country. But the farmers are still poor and marginalized despite that agriculture has been practised for many centuries in sub-Saharan Africa. This is attributable in part to lack of discernible and sustainable improvement in the productivity of smallholder agriculture. The farmers are poor not because they are lazy or small but because the traditional farming systems have low and declining productivity and there are no institutional, infrastructural and technological supports to ameliorate this situation (Owamo, 2003).

Food security and widespread malnutrition are about the greatest threats that sub-Saharan Africa faces despite that a larger proportion of the population is involved in agriculture. The vulnerability of the majority of African population to poverty and the low productivity of the agricultural system can be explained by, among others, the following factors:

Lack of consistent and sustained agricultural policy: There have been little efforts and commitment of successive governments to improve and protect smallholder agriculture. Policies have been formulated out of the socio-cultural context of this dominant group in the agricultural sector in favour of the few rich, large-scale, absentee farmers that produce plantation and cash crops and are rarely directly involved in the production of basic food crops. The overall effect is that the smallholder farmers continued to be marginalized and further disadvantaged.

Inappropriate technologies: Growth of the agricultural sector is highly dependent on the generation and dissemination of appropriate technologies that will facilitate higher productivity, higher income and reduction of production costs. Agricultural research in Africa has not addressed the constraints that the majority of smallholder farmers face under natural, rain-fed conditions, declining soil productivity and lack of basic inputs. Public agricultural research institutions show little interests in, and lack understanding of, the smallholders' situation. Much of the technology for smallholders is inadequate, largely because research priorities were not identified in consultation with farmers.

In the same vein, smallholder farmers have been encouraged to abandon traditional agricultural systems for a large-scale, intensive, mono-cropping system of agriculture (Longatu, 2003) without concurrent back-up supports, such as link roads, competitive markets and inputs that make such a system profitable. The mono-cropping system currently being advocated leads to the displacement and loss of a wide diversity of traditional landraces and crop varieties (Putter, 1994; Eyzaguire, 1995) on which generations of farmers have depended for feeding their families. Although these were not high yielding, they were well adapted to the local conditions and the diversity of plants farmed ensured that families would not go hungry and that they derived essential nutrients and vitamins need in a healthy diet.

Lack of access to productive land: This is a serious problem limiting the productivity of smallholder farmers, especially in West Africa (Owamo, 2003). Unlike the rich and stable volcanic soils of East and Southern Africa, soils in greater part of West Africa are fragile and easily degraded. But because virgin lands are no longer available, the limited land that each farmer cultivates continuously has been degraded and heavily 'mined' of essential nutrients. With the prevalent lack of an appropriate soil management system and the lack of access to fertiliser as supplement to enhance soil fertility, smallholder farmers cultivate the same land from year to year and this renders the soil less and less productive.

Non-availability of high quality seeds and an inappropriate seed system: Seeds used by farmers are often of low physiological quality and this reduces final yield (Louwaars and van Merriwjk, 1996). Traditionally, African smallholder farmers have produced their own seeds or exchanged seeds among each other. With the change of agricultural seeds into a commercial commodity, the traditional, informal seed system is systematically being replaced with the formal or commercial seed systems that leave farmers to the mercy of large, multinational seed production companies. But poor farmers lack the financial means to purchase expensive commercial or 'official' seeds. This system is not sustainable, because when they are willing and able to purchase these seeds, the seeds are either not available within their immediate environment at the time they need it or that the seeds available are of varieties that are not adapted to the farmers' environments (Ajayi, 2000). The formal system supplies about 10% of total seed requirements in most of sub-Saharan Africa (Mulioleka, 1999, Ajayi and Fakorede, 2003). Modern plant breeding often emphasises wide adaptation at the expense of seed quality attributes, for example adaptability to local conditions and production systems, suitability for on-farm seed saving by farmers for future planting (hybrids versus open-pollinated varieties), longevity in storage under harsh, ambient tropical conditions (Paliwal and Smith, 2002). Many studies have shown that farmers are aware of the availability of new and improved varieties of crops but they prefer and continue to use their low-quality seeds, because they have no guarantee of continuous supply of the improved seeds and of the seed's adaptability to their production system (Louwaars and van Merriwjk, 1996; personal communication with Professor MAB Fakorede, Obafemi Awolowo University, Ile-Ife, Nigeria, 1999). Similarly, the margin of yield differences between improved and farmers' varieties are often too low to be of any practical advantage for adoption (Tripp *et al.,* 1998; Wright *et* al., 1995).

Lack of incentives: Smallholders have very little, if any, access to credit facilities, trade information, competitive markets, feeder roads to transport produce from farms to markets, and basic social amenities all of which work to limit the potential of an average farmer to produce enough to feed his family and have enough to sell in order to be able to meet other needs.

Impact of agriculture on health improvement

There are many health problems plaguing the African continent and some of these problems are caused by food insecurity, poverty, malnutrition, diseases and general underdevelopment. Food insecurity and malnutrition lead to much human suffering, substantial productivity losses, diminished work performance and lowered cognitive ability. At the household level, food security refers to the ability of the household to secure, either from its own production or through purchases, adequate food for meeting the dietary needs of all members of the household. The nutritional status of each member of the household depends on several conditions among which are availability of food of sufficient varieties, quality and safety according to individual needs. Also, each family member must have good health status in order to benefit from the food consumed (FAO, 2003). The FAO also recognises that healthy, well-nourished people are both the outcome of successful social and economic development and constitute an essential input into the development process. Human health problems are highly correlated to food availability and choices and in a number of ways, health improvement can be achieved through the growth of, and improvement in, the agricultural sector.

Prior to the dispersal of plant seeds as the principal organ for regeneration, seeds are endowed with a lot of resources in the form of carbohydrates, proteins, fats and oils, minerals and other biochemical components needed for germination and survival upon shedding from the mother plant. These materials are in themselves the basic ingredients that man needs for survival as food for humans, feeds for livestock, and industrial raw materials. Grass seeds like those of wheat, rice, maize, barley, sorghum, provide the largest calories of human needs. Legume seeds/pulses are rich sources of protein. Seeds of rape plant, sunflower, sesame, and safflower are the main sources of oil. Much of the world's chocolate comes from cocoa seeds. Tea and coffee are all products directly derived from plant seeds. The seeds of cucurbits are particularly nutrient-dense as stores of protein, fibre, mineral elements and essential amino acids and leafy vegetables are rich sources of iron and Vitamin A (Anonymous, 2003).

In addition to the variety of foods, agriculture has been used as a long-term, sustainable option for improving health through the enhancement of the nutritional value of foods grown by farmers and a range of other agricultural products. Examples of this include the development of different 'value-added maize germplasm' such as quality protein maize (QPM), low-phytic acid and high oil varieties (Paliwal and Smith, 2002). As a way of fighting malnutrition in West Africa, the West African Rice Development Association (WARDA) developed the 'New Rice for Africa', NERICA (Kanayo et al., 2003) which, in addition to its high yielding capacity, has high protein content. Similarly, genetically engineered 'golden rice' with a high vitamin A content was developed to combat vitamin A deficiency-induced blindness, especially in Asia where rice is a basic staple (Shiva, 2000; Dave et al, 2002). Generally, protein contents of harvested plant parts, especially grains, have been enhanced with fertiliser application (Zhang et al, 1994). Nutritional value of table eggs has also been improved by varying the amounts of plant-based raw materials for layers' feed (Brown et al, 2000). Personal involvement in community-based rural and peri-urban projects has shown that by growing a few stands of different vegetables behind backyards, rural and middle-income households have access to fresh vegetables. This frees money that could otherwise have been used to buy these vegetables to meet other needs, including healthcare. Therefore, any improvement in agriculture is closely linked to improvement in the health and the general well being of the people whose livelihood depends directly on agriculture and those who consume agricultural produce, raw or processed.

Given these close links between agriculture, economic development, food security and health improvement, agricultural development efforts in sub-Saharan Africa require a holistic approach that recognises the peculiarity of the traditional values, systems as well as practices within each homogenous ethnic or regional group. Such an approach will integrate the totality of the lives of the smallholder farmers vis-à-vis the agricultural production system, economic empowerment, traditional knowledge and values of the people (Ajayi, 2001).

Bridging the gaps

From the above discussion, the development of agriculture has both immediate and long-term significance for the economic development of any country, particularly in sub-Saharan Africa. Therefore, the following sustainable strategies for bridging the gaps are advocated:

Sustainable and friendly agricultural policies: Sub-Saharan Africa is naturally endowed with a wide range of environments that permits the production of a wide range of plants- cereals, legumes, roots and tubers, fibres, vegetables, fruits, and by inference various products from these plants. Sustainable agricultural production, utilisation and conservation of these plants need to be stimulated through favourable, long-term, smallholder farmer-friendly policies. Agricultural policies and supports. The frequent disruption and changes in agricultural policies with changes in government discourage long-term planning and adjustment by farmers. These farmers need feeder roads to transport their produce to local markets. This will further facilitate direct contact to farmers by industries that utilise farm produce as raw materials. As a policy, local, agrobased industries need to be encouraged and protected with appropriate government policies. In turn, local industries should also be further encouraged to utilise locally available and produced raw materials as a way of creating markets for farmers. Importation of raw materials that can be sourced locally should be banned except when local production is not enough to meet demand. Smallholders also need to have unrestricted access to competitive local, regional and international markets. Importation of abundantly, locally available raw materials and foods, including eggs, meats and fruits in processed forms from outside the region is harmful to both local industries and agriculture in any country. Government patronage of locally made agricultural products will set a good example for the larger population to follow. Governments also need to keep an eye on agricultural research so that attention is given to the needs of smallholder farmers and to neglected minor crops that are of great importance to local food security.

The collective weight of individual governments must be put together in a synergistic way to develop and protect regional markets for agricultural produce and resultant agricultural products. To complement this, it will be necessary that some massive public awareness campaign be mounted to add value to what is grown locally and to convince people that many of local products can be more nutritious than imported ones. Such a collective effort will stabilise prices and competitiveness in regional and international markets.

It is the responsibility of governments to facilitate access of farmers to basic inputs and credit facilities. Small is beautiful, smallholder agriculture will be profitable, with or without subsidies, if legal and institutional policy supports are put in place. Farmers, through their community-based co-operative societies, need easy access to credit facilities in unconventional ways that do not exploit their vulnerability.

***Development and dissemination of appropriate technologies*:** Smallholder agriculture is the principal source of food and other agricultural products in Africa. Therefore, technologies that are aimed at improving agricultural production and rural economies must be targeted at the traditional smallholder farmers. Agricultural research must, as a matter of urgent importance, generate such technologies through participatory approaches that involve the farmers. The starting point is for agricultural scientists not to see themselves as think-tank for farmers but to recognise farmers as equal partners and stakeholders, not a bunch of helpless, illiterate group of people upon whom anything can be dumped in the name of improved technologies. Proceeding from this view point, agricultural research carried out in partnership with farmers beginning from the conceptualisation stage, needs to be involved in the following:

i. Land management: Access of individual smallholder farmer to new, long - term land holding for agricultural purposes is no longer possible in most parts of sub-Saharan Africa. Therefore, sustainable strategies for managing and maintaining the poor and fragile aggregate structure of most tropical soils and of reviving the poor fertility status of these soils need to be evolved. The use of farm waste and organic amendments needs to be encouraged. Similarly, the philosophy of 'greening' agricultural fields should be incorporated into traditional agriculture. This involves identification of extremely fast-growing, preferably leguminous plants that require minimal amount of water for growth and establishment within a short period and these plants could be incorporated into the farming systems to 'green' the fields rather than leave them bare at the end of a growing season. The plants would then be ploughed back into the soil while green thereby returning some amounts of nitrogen back into the soil for the following season. But such plants need to be carefully selected in such a way that their seeds could be broadcasted just before main crops are harvested without much additional labour. Also, suitable plants for 'greening' must be ones than will not become noxious weeds later.

ii. Promotion of the cultivation and utilisation of native food crops: Traditional agricultural systems need to be supported through productivity enhancing technologies and farmers need to be encouraged to continue to cultivate and use traditional food crops. This way, the genetic base of these crops will be maintained and health status of rural communities will be improved. This will be achieved through integration of leafy vegetables, diverse fruits and nuts all of which are rich sources of vitamins and minerals in addition to their medicinal properties, into the food culture of both urban and rural populations. What is often overlooked when encouraging farmers to drop traditional farming systems is that the existing infrastructures do not give an average rural farmer access to most of his and his household's needs that are not met through his production. This is worsened by the fact that within a community, farming practices are often homogenous and less diverse, a risk-aversion strategy that guarantees communal efforts to any problem through farmer-to-farmer exchange of ideas and community-based traditional knowledge. From the biodiversity point of view, there is often no concurrent active collection and conservation of indigenous food crops that are displaced when promoting a modern mono-cropping agricultural system.

iii. Post-harvest handling of crop produce: The climate in sub-Saharan Africa is harsh for the storage of both perishable and non-perishable crop produce. A significantly high proportion of farm produce and yields are quantitatively and qualitatively lost as a result of either lack of, or inappropriate technologies for, post-harvesting handling of crop produce. The smallholder farmer-producer bears the full consequences of this enormous loss that could have been sold to meet other needs. Research is needed in post-harvest handling and processing of crop produce in order to minimise losses during harvesting and storage. The overall effect of this will be an improvement in rural economies and an increase in the quality of life in the rural areas where the majority of the population reside.

iv. Development of appropriate seed systems and technologies: The central role of seed in agriculture has been recognised right from early beginnings. Seed is the only living external input in food production that responds to stimuli in quantifiable measures. Therefore, it sets the limit of response to other inputs, including fertilisers and other agronomic inputs. The level of harvest obtained at the end of a season depends on the quality of seed planted (Ajayi et al, 2001). From an agronomic perspective, quality encompasses physical, biological, pathological and genetic seed parameters that ultimately have a bearing on plant performance and final yield of a crop in the farmers' fields (Hampton, 2002). There is a high level of awareness about seed quality among all stakeholders, including farmers and policy makers. Farmers do not just use any germinable material; instead, different criteria and combinations of physical characteristics are employed to 'select' seeds that show the potential of producing good crop. These selections are then carefully handled, thus underscoring traditional farmers' recognition of 1) the role of seed as a biological and an agricultural input and 2) the need to preserve its quality in order to get a good crop from it. In Mexico for example, farmers have recognised that quality and field performance of flat maize seeds is better to that of round; consequently, ratio of the prices of flat to round seeds could be as high as 1:0.6 (Batistella *et al, 2002).*

Information on traditional seed practices is scanty. A thorough understanding of the traditional seed systems is needed in order to identify the weak points where there is room for improvements in partnership with the farmers themselves. In contrast to the commercial or formal seed systems that have failed smallholder African farmers (Almekinders *et al, 1994),* community based, informal seed systems and approaches to seed production and distribution need to be promoted and supported. But their incorporation into traditional agricultural systems and sustainability needs thorough scientific and anthropological investigations. Some of the steps involved in post harvest handling of seeds in commercial seed production are more damaging than beneficial to seed quality and confer no yield advantage, irrespective of the agricultural system in which the seeds are planted (Ajayi, *2003;* Ajayi *et al, 2002).* Unfortunately, the cost of these operations is passed to farmers. To be sustainable, however, community-based seed systems must guarantee farmers' access to their own varieties and to other improved adapted varieties and in partnership with community-based seed growers, quality control measures must be built into the system.

v. Tailor-made extension system: There is need for a new extension system, one in which farmers are the ultimate beneficiaries. Past systems failed because they were imported into a socio-cultural context that was different from the ones for which they were developed. To help farmers, agricultural extension experts first need to know what the perceptions and priorities of the farmers are with respect to the constraints that they face. As in the case of seed systems discussed above, understanding traditional farmers' perception requires a change in the approach of extension agents so that farmers are elevated from the position of poor beneficiaries of extension services to that of intelligent collaborators that we need to work with to improve agriculture.

The nature of agricultural training also contributes to the failure of both research and extension in helping farmers. A single system of delivering extension services to farmers has failed largely because traditional farming systems are deeply rooted in the culture of each community and are thus home-grown and community-farmer based. Therefore, the sociological content agents and managers need to be increased generally, a new holistic agricultural training paradigm is imperative so that the gap between the theoretical and applied aspects is bridged. In the transition stage, basic agricultural research is necessary only to the extent that it helps to find solutions that can be applied to practical problems of the smallholder farmers.

References

Ajayi, S. A. 2000. "Challenges of Private Seed Sector Participation in Agricultural Development in Africa." *West Africa Seed and Planting Material Newsletter* 6: 2-3.

Ajayi, S. A. 2003, *Physiological and Biochemical Basis of Maize Seed Quality.* Cuvillier Verlag, Göttingen.

Ajayi, S. A. and M. A. B Fakorede. 2003. "Nigerian Seed Industry." *International Society of Seed Technologists Reports.* 2(4): 21-26.

Ajayi, S. A., M. A. B. Fakorede, G. Rühl, and J. M.Greef. 2001. "Defining Seed Quality by Seed Maturity and Crop Performance: Observations on Tropical Maize." *Journal of New Seeds* 3(2): 49-71.

Ajayi, S. A., Rühl, G. and J. M. Greef. 2002. "Interactive Effects of Mechanical Harvesting and Processing, Storage Conditions and Duration on Quality of Maize Hybrid Seeds." *Mitteilungen der Gesellschaft für Pflanzenbauwissenschaften* 14: 177-178.

Almekinders, C. J. M., N. P. Louwaars, and G. H de Bruijn. 1994. "Local Seed Systems and their Importance for Improved Seed Supply in Developing Countries." *Euphytica* 78: 207-216.

Anonymous. 2003. "Pumpkin (cucurbita) Seeds." Online, available http:// www.wholehealthmd.com/refshelf/substancesview/1 525,10113,OO.html. Retrieved August 2003.

Barret, C. P., F. Place, and A. A. Aboudi (eds.). 2002. *Natural Resources Management in African Agriculture: Understanding and Improving Current Practices.* CAB Int., Wallingford.

Batistella, F. F., F. V. Möro, and N. M. Carvalho de. 2002. "Relationships Between Physical, Morphological and Physiological Characteristics of Seeds Developed at Different Positions of the Ear of Two Maize *(Zea mays L.)* Hybrids. *Seed Science and Technology* 30: 97-106.

Brown, A. A., S. A. Ajayi, and S. A. Fala. 2000. "Effect of Supplementation of Layers' Diet with Mixtures of Different Pepper Species on Egg Quality." *Nigerian Journal of Animal Production* 27(1): 71-76.

Dave, D. R. Robertson and L. Unnevehr. 2002. "Golden Rice: What Role could it Play in Alleviation of Vitamin A Deficiency?" *Food Policy* 27(5/6): 541-546.

Eyzaguire, P. B.. 1995. "*In situ* Conservation and Sustainable Use of Minor Vegetable and Fruit Species." In Engels J. M. M. (ed.) *In situ Conservation and Sustainable Use of Plant Genetic Resources for Food and Agriculture in Developing Countries*. IPGRI, Rome/DSE, Feldafing. pp 104-106.

FAO. 2003. "Household Food Security and Community Nutrition." Online available http://www.fao.ora/es/ESN/nutrition/household en.stm. Retrieved August 2003.

Hampton, J. G. 2002. "What is Seed Quality?" *Seed Science and Technology* 30: 110.

Kanayo, F. N., P. J. Kouka, and M. T. Jones. 2003. "South-south Cooperation of Food Security in West Africa." Online, available http://www.ec.-dc.net.cn/ events/cosouth200102/9.htm. Retrieved July, 2003.

Longatu, S. R. 2003. "Multi-agency Partnership for Technical Change in West African Agriculture: Nigeria Case Study Report on Rice Production." Ecosystems Development Organisation, Jos.

Louwaars, N. P. and G. A. M. van Merriwjk. 1996. *Seed Supply Systems in Developing Countries,* Wageningen: CTA.

Mulioleka, S. W.. 1999. The Challenge of Seed Production and Supply in Africa." *Seed Science and Technology* 27: 811-824.

Owamo S.W. 2003. "Policy Research on African Agriculture: Trends, Gaps and Challenges." ISNAR Research Report 21. The Hague: International Service for National Agricultural Research.

Paliwal, R. L. and M. E. Smith. 2002. "Tropical Maize: Innovative Approaches for Sustainable Productivity and Production Increases." *Crop Improvement-Challenges in the Twenty-first Century,* (Ed. M.S. Kang), pp 43-73, The Haworth Press, Inc. New York.

Putter, A. 1994. *Safeguarding the Genetic Basis of Africa's Traditional Crops.* CTA, The Netherlands/IPGRI, Rome.

Shiva, V. 2000. Genetically Engineered Vitamin 'A' Rice: A Blind Approach to Blindness Prevention. February 2000. Online available www.biotech-info.net/blind rice.html. Retrieved August 2003.

Tripp R., D. J. Walker, A. Opoku-Apau, A. A. Dankyi, and L.L. Delmini. 1998. "Seed Management by Small-scale Farmers in Ghana. A Study of Maize and Cowpea Seed in Brong-Ahafu and Volta Regions," *NRI Bulletin* 68, Natural Resources Institute, Chatham, UK.

Wright M., L. Delmini, J. Luhanga, C. Mushi, and H. Tsini. 1995. *The Quality of Farmer Saved Seed in Ghana, Malawi and Tanzania.* Natural Resources Institute, Chatham, UK.

Zhang, F., A. F. Mackenzie, and D. L. Smith. 1994. "Nitrogen Fertiliser and Protein, Lipid, and Non-structural Carbohydrate Concentrations during the Course of Maize Kernel Filling. *Journal of Agronomy and Crop Science* 172: 171-181.

THE DEVELOPMENT OF AGRICULTURE AND ITS IMPACT ON HEALTH IMPROVEMENT IN SUB-SAHARAN AFRICA

Uche C. Amalu

Summary

The early concepts of the origin of agriculture attributed the development of agriculture to the "discovery" by the hunter-gatherers of the pleistocene age that planted seeds would sprout. As soon as the process of planting, followed by the domestication of some wild animals was discovered, the hunter-gatherers adopted them as a basis of their subsistence. However, the gradual development of agriculture holds far-reaching positive and negative consequences for mankind. Meeting the demand of the ever-growing populations for agricultural products can only come from growing more food crops and livestock. The sub-Saharan region of Africa has the potential to be transformed, if not into a Garden of Eden, at least into a cornucopia of food, sufficient to give all its inhabitants an adequate diet: call it self-food sufficiency; call it food security. But growing our food is often bad for our health, just as poor health hampers us in growing our food. First, accidents in the field and on the farm require as much attention as the appalling lack of safety elsewhere in the food chain. For procedures, processors and consumers world wide, safety measures are often too little, too late and - for the unorganised - too far away. Secondly, we can feel the pain in food production each time a fever aches its way through our bodies. Diseases such as the Acquired Immunodeficiency Syndrome (AIDS), Malaria, River blindness, Guinea-worm disease Schistosomiasis, Lymphatic filariasis Diarrhea and Respiratory diseases, all have direct and negative impact on many aspects of our lives - including agricultural production. It is necessary to understand the forces, which are driving the diseases epidemic, or rather pandemic, now that it is a global problem. Sustainable development of agriculture and the management of resultant health conditions in the sub-Saharan Africa will depend mainly on changes of behaviours and attitudes on the part of the people. The same applies to coping with the permanent impact of disease conditions on agriculture and rural development in sub-Saharan Africa. The disease prevention and control strategies stress the selective use of preventive measures wherever they can lead to sustainable results. Such measures should be aimed at halting the deterioration of the disease situation, contributing approximately to the development of health services, and encouraging inter-sector (agriculture and health) co-operation and community participation.

Partnerships, networking, and the promotion of equal opportunities in endemic communities will always constitute the core of disease control and prevention strategies. All the stakeholders, again working in collaboration should lead in concert in the analysis of past and current agricultural development programmes in sub-Saharan Africa and their impacts on the health of citizens. This should enable experts to identify causes of health problems associated with agricultural development, including zoonotic diseases and new products of biotechnology, in the region and develop appropriate interventions to reverse or modify underlying factors and define or re-define future actions.

Introduction

The early concepts of the origin of agriculture attributed the development of agriculture to the "discovery" by the hunter-gatherers of the pleistocene age that planted seeds would sprout and that some wild life can be reared. The lives of the pre-agricultural savages were up till then characterised as "nasty, brutish and short". As soon as the process of planting, followed by growth of such plants was discovered, the hunter-gatherers would have been quick to see plant cultivation as a major advance and to adopt it as a basis of their subsistence (Mine and Vandermeer, 1990). Underlying this view were two assumptions concerning the relative advantages of agriculture. First, agriculture was a more reliable and more productive form of subsistence than hunting and gathering. Secondly, an agricultural way of life required the devotion of less time and energy to subsistence pursuits, thereby leaving more leisure time for the development of those arts, crafts and technologies, which characterised higher civilisations.

The work of anthropologists among current hunter-gatherers, however, has discounted this simplistic model. They argued that there were not known hunter-gatherer groups so primitive that they did not appropriate the changes in growth of plants from sprouting to fruiting. On the contrary, these people had a very sophisticated knowledge of their ecological surroundings. Further, ethnographic studies of the hunter-gatherer groups have more broadly demonstrated that the life of these people was not one of constant struggle with nature. In contrast, it is highly unlikely that early agriculture involved so little effort or provided so much security. Although farming production conditions generated variable crop yields and the stored harvests were vulnerable to attack by pests, moulds, theft and fire; settled life brought with it the higher risks of dirtiness and disease infestation, while fixed farm sites increased the prevalence of crop pests and weeds. Worse, having now settled for an agricultural life, early farmers lost the flexibility and mobility of hunter-gatherers to pursue alternative sources of livelihood. One of the attendant problems was the increased restriction of the habitat of wild plants and animals, making these sources of food increasingly unavailable and thereby creating food shortages.

Given the relative ease of the hunter-gatherer subsistence in comparison with agricultural subsistence, a curious person would ask the question: why did agriculture evolve at all? So many schools of thought have proffered reasons. One school posited that some sort of pressure was necessary to compel the hunter-gatherer groups from the wandering (hunting and gathering) type to the settled (agriculture) mode of subsistence. Frequently this "push" is envisioned as an imbalance between people and resources that would make the advantages of agricultural systems (higher productivity per unit land area) outweigh the disadvantages (higher demands of time and energy) of hunting and gathering systems. Yet another school posited reasons from some possible factors creating population-resource imbalance such as (a) environmental changes leading to a reduction in the density of available wild resources and (b) increases in population to a level that the carrying capacity of the environment became stressed. In both cases, however, the approaches have been basically the same - deterministic; that is, a radical change in the environment or population leading to radical change in subsistence practices (Mine and Vandermeer, 1990).

The transition from food collection to food and livestock production holds far-reaching consequences for both present-day ecological and societal interactions. Within the natural environment, the role of farmers changed dramatically from participant to manipulator status, while farm sizes changed extensively from small pieces to large-scale acreages. Within the social realm, equally dramatic changes marked the transition from simple, egalitarian societies to the present day political, economic and social hierarchies of higher civilisations. It is in this context, that studies of the development of agriculture and its impact on improvement of human health should now and in future form very important aspects of our research agenda in sub-Saharan Africa.

Growing human population growing more food and health implications

The population of the developing world has more than doubled since 1965 and now stands at over five billion. By the end of the decade it is estimated that the population will have grown to six billion. This growth in human numbers has been a principal cause of a rising demand for food, water, and other life-sustaining resources in the past and will continue to be so for the foreseeable future. The United Nations projects that the developing-country population will reach 6.5 billion by 2020 and 8.2 billion by 2050 (Bongaats and Bruce, 2001). Elsewhere, the United Nations provides three possible projections of world population, ranging from 7.7 billion for 2020 through 9.4 billion for 2050 to 17.5 billion people in the year 2100 (Wilson, 2001). Almost three quarters of the world's population live in agricultural societies geared towards survival, relying on local raw material and powered by man and animals. It is within this section of the society, particularly the sub-Saharan Africa, that the population growth will be greatest.

Across much of sub-Saharan Africa, however, though modest gains in agricultural production have been achieved the gains have not kept pace with population growth and the migration of people from the rural areas to the cities. It has been estimated that even if annual population growth were to stabilise at 2.5 percent, the demand for agricultural products in sub-Saharan Africa by the year 2025 would be three times greater than it was in 1990 (IITA, 1993). Meeting the demand for agricultural products can only come from growing more food. The sub-Saharan region of Africa has the potential to be transformed, if not into a Garden of Eden, at least into a cornucopia of food sufficient to give all its inhabitants an adequate diet: call it self-food sufficiency; call it food security. But growing our food is often bad for our health, just as poor health hampers us in growing our food. First, accidents in the field and on the farm require as much attention as the appalling lack of safety elsewhere in the food chain. For producers, processors and consumers world wide, safety measures are often too little, too late and - for the unorganised - too far away. Secondly, we can feel the pain in food production each time a fever aches its way through our bodies. Diseases such as the Acquired Immunodeficiency Syndrome (AIDS), Malaria, River blindness, Guinea-worm disease, Schistosomiasis, Lymphatic tilariasis, Diarrhoea and Respiratory diseases, all have direct and negative impact on many aspects of our lives - including agricultural production. We all need to eat, and we all need to live. Agriculture is what makes us what we are. In the face of challenges such as the need to eat, the need to live, the accidents and the negative impacts of agricultural development, there usually is a sting of a dilemma. The fact remains that we all need to eat, and we all need to be safe and healthy

Rural agriculture and health

In sub-Saharan Africa, food production is considered to be the most hazardous 'sector' after mining and construction. About half of the world's 250 million workplace injuries are estimated to occur in agriculture. More than one million people die each year through work-related accidents and sicknesses in the agricultural sector, according to new statistics from the International Labour Organisation (SPORE, 2002). Previous figures (SPORE, 2001) indicated that more than half of the world's 1,700,000 occupational deaths were in agriculture. Now, it is the same share of a total of 2,000,000. The sector employs one-half of the world's labour force, with about 1.3 billion workers active in agricultural production between 10 and 15 percent of these deaths take place in sub-Saharan Africa.

The foregoing statistics show that rural agriculture is not a romance at all. Among other things that it means to us it is a rough, tough world of injury, disability, disease and death. For thousands of years, rural agriculture has fed many, and it continues to support the livelihoods of 300 million people world-wide. And that is a conservative estimate. There are overwhelming numbers of farming methods and materials that constitute health hazards for the present and future agriculture workers. These include the use of traditional tillage tools, (Indian hoes, ox-driven sticks, etc.) sharp iron and wooden planting tools (knives, sticks and bamboo sticks), high drudgery harvesting and processing tools (sickles, knives, baskets, threshers, etc.) and untreated seeds. Others include the uncertainties of contracting illnesses from our livestock, unique pests and disease-carrying vectors/pathogens, disease-carrying soil and water bodies and outmoded post-harvest technologies. For instance, in rice crop harvesting, manual method is still prevalent in Africa. "The sickle is commonly used to cut the stems near the base of rice plants. Several manual methods of threshing are common. The panicles of cut rice plants are beaten against bamboo slabs or on hard wood over a wide mat to catch the flying grains; they are beaten by flail, rubbed with the feet; or fed onto drum threshers. Most women still use the hoc as the major tool for farming in spite of their physiological and anthropometric limitations. Their low capital base and inability to obtain credit keep them tied to hoes. On their own part, men still operate some dangerous mechanical tools manufactured to ease work for agricultural workers.

What the entire foregoing discourse means is that all along the food chain, from plough to the plate, there is constant exposure of the farmer to health hazards and usually there is little or no respect at all for the safety of those who produce and process the food. Their health and safety are vital and not only for themselves, but also for the entire nation; after all, without a safe production process, without safe producers there cannot be safe consumers. The agricultural labour force in the African states and other developing countries is known for its high levels of self-employment, although overall labour is in some form of waged unemployment. Self-employed or not, no worker deliberately injures himself or herself, although many people will take highly dangerous short-cuts in their work, under pressure to produce and earn marginal incomes. But danger there is, partly rooted in a lack of safety awareness and education, partly in the lack of complete training and instructions, and in part in the age-old predominance of gain over sane.

Urban agriculture and health

More people are moving to towns than to rural areas. In Africa, 35% of the continent's population of 749 million in 1998 is urban, and it is growing at a rapid 8% each year. To face this challenge, conventional thinking strives to "feed the city" by increasing production faster than population growth (SPORE, 1999a). In recent times, urban agriculture has won the attention of agriculturists and policy-makers alike. All over the world, case studies are sprouting like mushrooms in a stairwell grower. They describe how rich and poor urbanities develop new farming systems based on small livestock (rabbits, goats, guinea pigs, pigs, etc.); poultry (chicken, ducks, guinea foul, etc.), vegetables, fruits, and herbs. They also discuss forestry, which yields fruits, nuts, fuel-wood, and building materials. Urban agriculture is described by some as a more significant approach to development than innovations in the informal economy, bartering, or micro-finance.

The expansion of urban agriculture in Africa has been stimulated by an economic structural adjustment program and the collapse of much of the formal economy. Most urban agriculture is small-scale in many countries and is dominated by women. For instance, in Calabar and Uyo cities of far south-east Nigeria, urban agriculture improves the lives of women not only economically, through income-generation and substitution of purchases, but also nutritionally, through greater variety and freshness of food. It also increases food supply because of higher yields of herbs and vegetables on composted areas and from hydroponics (the process of growing plants in water with added nutrients). The major exceptions to women-dominated urban agriculture are Brazil, China, and Senegal, where men dominate. There are environmental benefits, too: more vegetation improves air quality by acting as a sink for gas emissions from fossil fuels and through the cooling effect of evaporation. Compost, much used in urban agriculture, facilitates waste management.

There are health hazards in urban agriculture however, and these call for attention. A leading proponent, Jac Smit of the Urban Agriculture Network (UAN) stresses the danger of uncontrolled food production in cities. There are risks of contamination from untreated human and animal waste in compost, and from auto-sump oil and heavy metals; tuberculosis from cattle; insecticide poisoning; plague spread by rats attracted to compost bins, groundwater and soil pollution from agricultural chemical, and malaria from mosquitoes breeding in unmanaged irrigation pools. Nitrate is currently recognised as the most serious agricultural chemical pollutant. As nitrate is detected in drinking water (usually source from ground or surface waters), it causes methemoglobinemia in children. Agriculture contributes to ground-water nitrate through the leaching of nitrogen fertiliser, animal wastes and crop residues. Phosphorus and nitrates are the major contributors to eutrophication in aquatic systems and aquatic life. The dangers of accumulation of pesticides, especially organochlorines, in aquatic ecosystems are clear. These chemicals, although not highly toxic at low concentrations, stay in the bodies of aquatic consumers and eventually enter and accumulate in the food chain.

Working the soils and human health

Soil, the baseline for producing food both for man and his livestock, can affect human health adversely in several ways leading either to specific diseases or to more general ill health. Locally it emits radon, which is radioactive and a carcinogen. Radon from poorly drained soil has been linked recently with infant mortality. Elements from the soil pass through the food chain of plants and animals and also into the water supply system. They can occur in concentrations that are either toxic or deficient, both leading to ill health. In general, it is the trace elements (also called heavy metals) that are important and many of which are considered essential, although they can all be toxic if their concentrations exceed certain limits. These elements include aluminium, arsenic, cadmium, copper, fluorine, iodine, lead, selenium, thallium and zinc. Some of the effects of soil on health are due to its natural condition, others are exacerbated or due entirely to man's activities, especially pollution. For some elements there are clear and direct causes of diseases. Examples include *Keshan* disease caused by selenium deficiency, and *Itai-itai* disease caused by excess cadmium. They are especially evident in isolated communities, which derive most of their food and water locally. People in them are more likely to suffer from poor food than are those in the towns and cities and in the developed countries where food comes from widely separated areas (Mills, 1987). Varied sources of food provide considerable protection against diseases associated with the toxicity or deficiency of elements. In the developed countries it is more difficult to detect relations between soil and health. Nevertheless, people in the developed countries might be less healthy than they should be because of ignorance about factors in their diet.

The effects of soil on human health have three main facets, namely (1) the constituents and behaviour of the soil itself, (2) the pathways by which the constituents move from the soil into the human body, and (3) the mechanism by which the constituents once in the body affect health. Soil scientists as a whole are familiar with the first. The movement of elements through the food chain is well known quantitatively, as are the life cycles of certain parasites. The ingestion and inhalation of other harmful constituents from the soil are less appreciated in general, although there are several well-documented case studies. The effects of certain parasites and some soil-borne pathogens in the human body are known reasonably well. However, the effects of most of the trace metals seem to be less well understood, partly because of the interactions between them, and partly because of the complexity of our bodies and of the metabolic reactions in them. When we combine these three aspects and try to relate the causes of either thrift or illness in people to the soil on which they depend, we realise that the links are remarkably poorly understood.

Injuries, dangers and diseases in agricultural work

Generally, the high levels of fatalities and injuries in agriculture - twice as high as in most other sectors - are due to its special conditions. Much agricultural work involves multiple tasks and multiple locations, most of them in the open air, exposing workers to climatic conditions. Much work is seasonal, with added urgency. Workers are in frequent or constant contact with animals and plants, and are exposed to bites, poisoning, infections, parasitic diseases, allergies and toxicity. The postures required in performing agricultural tasks, coupled with ill-designed manual tools or mechanical equipment, lead to injury and disabilities.

The levels of exposure to pesticides and other chemicals are renowned - many small-scale farmers are unaware of the dangers of pesticides and have no practical access to either clear instructions or protective clothing. The safety position in formal large-scale farms is often no better. There are many cases in some countries of sub-Saharan Africa where neither training nor protection is available to most workers, at least not the casual temporary workers (Spore, 2001). Away from the field, the lack of safety is equally alarming. Small ill-lit, ill-ventilated workplaces with poorly protected equipment are often the icons of agro-processing plants, at least in the informal sector. The risks for the safety of workers, and the hygiene of the end product, are serious. As production intensifies, the techniques for crop cultivation and livestock rearing, food processing and distribution become more complex. The introduction of more machinery and numerous chemical compounds, from fertilisers to cleaning agents, each brings potential harmful humans and environmental effects in their wake.

Many agricultural practices have the effect of encouraging breeding grounds for disease vectors causing a real clash between the interests of food producers and health. Areas become epidemic when conditions that normally limit transmission change radically as a result of abnormally heavy rains, long periods of increased humidity and temperature or more permanent changes of microclimate due to development of irrigation systems agricultural projects or tree plantations. The large-scale dams built for water and energy supply and the small dams and fish ponds currently favoured for irrigation and protein production are both homes to insect larvae and other vectors of diseases. Similarly, the avoidance of pesticides in 'sound agricultural practice' has encouraged more insect population. Even health and hygiene and commercial advances in packaging have their knock-on effects on health. Cleaner food, yes, but the packaging, when discarded, adds to the breeding opportunities for insects.

Animal husbandry and human health

The domestication of animals and intensification of livestock production brought with them increased interactions between humans and animals and the attendant expansion of zoonoses and other communicable diseases common to humans and animals. Ever since, emerging and re-emerging zoonotic diseases have acquired both regional and global significance. These diseases include salmonella enteritidis in poultry, the most frequently reported zoonotic disease in many countries, multi-drug resistant salmonella typhimurium, Marburg and Ebola viral haemorrhagic fevers in Africa, Rift valley fever in cast Africa, the Arabian Peninsula and Egypt, the New World screw worm (Cochliomyia hominivorax) in north Africa and new rabies-like viruses in bats in Australia and Europe. The unexpected link between bovine spongiform encephalopathy (BSE) and variant Creutzfeldt-Jakob disease (vCJD) called for close inter-sectoral co-operation to elucidate this connection. Resistance to antimicrobials among zoonotic bacteria has also become an issue of increasing concern for animal production and human health. In recent years, the threat of a global influenza pandemic has resulted in renewed research actions in relation to mammalian and avian reservoirs. There are many reasons for the increased occurrence of zoonotic diseases, including alteration of the environment, establishment of human settlements in formerly uninhabited areas, a greater demand for animal protein, intensification of animal production, and acceleration of trade in live animals, animal products and other foodstuffs. Food animals are the reservoir for many emerging and important food borne diseases, including those caused by Escherichia coli 0157:H7, non-typhoidal salmonella spp., campylobacter spp. and Yersinia spp. Numerous emerging; and re-emerging diseases are directly transmitted from animals to humans; in some cases, the animals act as intermediate or accidental hosts, while in others transmission occurs via arthropod vectors. The veterinary public health programmers need to emphasise the importance of prevention of such diseases by highlighting their economic and health consequences.

Agricultural biotechnology and health

Agricultural biotechnology is viewed as the most important technological tool that will shape world agriculture over the coming decades. This agricultural tool offers prospects for further growth in productivity, saving raw materials, reducing energy consumption, and mitigating pollution problems, and many more. However, several questions concerning the new biotechnology arise, including the following: will the introduction of biotechnology result in a continued rapid growth in agricultural productivity? How will technology transfer take place between and among the developed and developing nations. What are the health implications of the biotechnological tools and products? A major issue that will affect successful applications of biotechnology to agriculture is the regulatory climate governing the release of the new products. A safe and efficient regulatory process, able to ensure public health and environmental safety, is in itself a comparative advantage in biotechnology. However, both food safety and bio-safety regulations should reflect international agreements and a society's acceptable risk levels, including the risks associated with not using modern biotechnology to achieve desired goals.

Some diseases and their effects on farmers' health and agriculture

Of the six major communicable diseases, killing more than a million people a year, one, the HIV/AIDS, is transmitted by direct bodily contact; it is the leading killer disease, its effect on agriculture and rural life too sadly evident in our daily lives. However, also of special relevance to agriculture is Malaria, and its vector, or carrier, the mosquito, guilty of two million deaths in Africa alone each year. River blindness (Onchocerciasis) is yet a major example of disease transmitted by nematodes carried by such external agents as black-fly. Other diseases of immense adverse importance include Guinea worm infection (Dracunculiasis or Dracontiasis), and Schistosomiasis (Bilharzias). Yet diarrheal diseases and acute respiratory diseases, such as cholera, giardiasis, influenza, pneumonia or whooping cough, thrive in damp and unclean environments, and siege on the ill-nourished young and old. The same conditions encourage the spread of tuberculosis now gaining ground in many countries as the number one killer of adults, including many HIV/AIDS sufferers.

HIV/AIDS disease

The talk of Human immunodeficiency virus and acquired immuno-deficiency syndrome (HIV/AIDS), the sexually transmitted disease that strip the body of its immunity to infections is that of devastation, decimation and despair. Perhaps this is why the "Red Ribbon" has become a worldwide symbol for mourning a death from AIDS, a disease whose impact on African, Caribbean and Pacific countries agriculture is terrifying for those who suffer and die, those left behind, and those yet to come (SPORE, 1999). Slightly more than 20 years ago, the first cases of AIDS were identified. Since then, scientists not only have identified the human immunodeficiency virus that causes AIDS, but also understand many of the stages in transmission. Naturally, an infection that is transmitted in a predominantly heterosexual manner and that destroys the health and finally the lives of people of prime producing age exerts a considerable socio-economic impact (Barnett and Rugalema, 2001).

Today, approximately 36 million individuals are living with HIV/AIDS world-wide. Assuming that each HIV/AIDS case directly influences the lives of four other individuals, the disease is affecting a total of more than 150 million people. More than two-thirds of all new AIDS cases in the world are reported in Africa, where more than 21 million people suffer. The regional HIV/AIDS statistics and features at the end of year 2000 (Ninja, 2001) are given in Table l. Unlike many diseases, AIDS affects individuals in their most productive years, with the majority of HIV infections occurring in people between the age of 20 and 40. Most, if not all, of the 25 million people in sub-Saharan Africa who are living with HIV/AIDS will have died by the year 2020, in addition to the 13,7 million Africans already claimed by the epidemic. In some rural communities in sub-Saharan Africa, this selective nature of the disease has practically wiped out a whole generation, leaving behind the grandparents and the orphaned children to face the challenge of survival (Fleshman, 2001). It is for most people a disease of poverty, although it can strike anyone indulging in unsafe sex and - greatest concern of all - the unborn of sufferers.

According to Barnett and Rugalema, (2001) neither a vaccine nor a cheap, assured, and effective treatment for HIV/AIDS exists. For this reason, the pandemic continues to grow and to affect millions of people worldwide, particularly the poor in the southern hemisphere, where 95% of cases are concentrated. With most illness and death occurring in people of ages 15-50, the disease deprives countries, communities, and households of their strongest, most productive people. When HIV/AIDS strikes, it strips away assets of all forms - human, financial, social, physical, and natural (Table 2), including its effect on the ability of households to acquire enough nutritious food for members to lead active, healthy lives; and exorbitant health-care costs, labour shortages, declining asset base, breakdown of social bonds, downgraded crops, and loss of livestock. All of these effects lead and contribute to food insecurity.

HIV/AIDS morbidity and mortality affect food security by reducing households' ability to produce and buy food, by depleting assets, and by reducing the insurance value of social networks as household calls in favours. AIDS mortality permanently removes adult labour from the household. Medical costs associated with caring for the sick and bedridden have to be borne, along with the funeral expenses of family members who die of the disease. This combination of adult morbidity and mortality and the associated reallocation and withdrawal of labour has led to a number of changes: (a) downgraded crops and loss of livestock leading to chronic food insecurity, (b) loss of farm management resources and skills, braking down the chain of knowledge transfer and labour sharing between generations, (c) inability to earn income from non-farm activities, forcing desperate measures to be taken including taking unaffected children out of school to save expenditure on fees, books and uniforms and to provide extra labour, (d) loss of assets such as outright sale of livestock, family land and household property to generate ready cash for food and to defray large health and funeral expenses, (e) disruption of social values and networks due to disinheritance, discrimination and stigma, and increasing dependency of the debilitated, disabled and the unaffected. These measures can help sustain families for a while but, in the long term, they worsen the prospect of being able to earn a living and ensure an adequate supply of food. The threat to household and community food security is a serious one, particularly as a good nutritious food diet has been shown to improve the condition of those living with AIDS (Spore, 2000). Finally, the overall strength of communal and associational life of community dwellers, mutual interpersonal trust, and norms of reciprocity may be weakened as existing social values and networks ostracise people stigmatised by the disease.

Malaria

At present, about 100 countries or territories in the world are considered malarious, almost half which are in Africa, south of the Sahara (Fig. 1a) Although this number is considerably less than it was in the mid-1950s (140 countries or territories) more than 2400 million of the world's population are still at risk (WHO, 2000). The incidence of malaria world-wide is estimated to the 300-500 million clinical cases each year, with about 90% of these occurring in Africa, south of the Sahara. Malaria is caused by infection by any of four species of Plasmodium but most of the malaria incidences are caused by Plasmodium falciparum. Plasmodium falciparum causes falciparum (malignant tertian or sub-tertian) malaria which is the most serious form of malaria and can be rapidly fatal in non-immune individuals if not treated promptly; Plasmodium vivax causes vivax (benign tertian) malaria which is widespread but rarely fatal, although symptoms during the primary attack can be severe; *Plasmodium malariae* causes quartan malaria which is generally mild but can cause fatal nephrosis; and *Plasmodium ovale* causes ovale (ovale tertian) malaria which is the least common type of malaria, and produces clinical features similar to *Plasmodium vivax* (Anon, 1996).

The public health problems associated with malaria are noticeable world-wide but more so in African rural areas, where malaria frequently strikes at the time of the year when there is usually the greatest need for agricultural labour and during harvests. Malaria, either alone or in combination with other diseases, is estimated to kill between 1.1 and 2.7 million people world-wide each year. Of this number, approximately 1 million children under age 5 in sub-Saharan Africa die of malaria each year. Malaria causes about 25 percent of all deaths among children in the region, especially among children living in remote rural areas with poor access to health services. Malaria also contributes to low birth, one of the leading risk factors for infant mortality, because pregnant women are more susceptible to both malaria and anaemia. Malaria has also had severe and, in some instances, devastating effects on adults engaged primarily in agriculture, extension and veterinary services, and consequently in some sub-Saharan African countries there are insufficient staff strengths to maintain these services. These invariably affect the timely reporting of, and reaction to crop failures, crop losses in the field or storage and livestock disease outbreaks. It is, perhaps, no great surprise that in Zambia, Uganda and Tanzania, there has been a significant decline in crop yields in small farms, especially over the last five years. This decrease in agricultural output can in part he blamed on the multiple effects of the malaria epidemic. Illness and deaths directly affect the availability of farm labour, and those who remain healthy are preoccupied with the problem of caring for the sick or undertaking additional tasks that were previously done by others. As a result, time-consuming works like weeding, pruning, mulching and land clearance is either not carried out properly or is left undone. Extension services, already over-stretched and under-resourced, suffer from absenteeism due to long illnesses, attendance at funerals, and death of workers. Failing to carry out essential agricultural practices also causes a decline in soil fertility, an increase in pest and disease incidences and an inevitable reduction in overall output. In some regions of Ghana, Zimbabwe, and Uganda, for example, more than 8% of the extension workers have died of malaria. By cutting down large slices of today's and tomorrow's productive generations, the malaria disease aggravate existing agricultural and rural problems. Rural women suffer the most from the impact. For the women, it is the situation of triple jeopardy arising from their rootes as mothers, wives, and farmers. When they have to replace the loss of a child or breadwinner by starting or expanding their agricultural activity, they are hindered by uncertainty of access to land, credit and education.

The importance of malaria in sub-Sahara Africa can never be over-emphasised. According to the World Health Organisation (WHO, 2000) the single disease of malaria reduces the gross national product (GNP) of countries in sub-Saharan Africa by more than 10%, rising to as much as 2-6% in Kenya and 1-5% in Nigeria. The estimated annual direct and indirect cost of malaria in Africa alone is more than USD 2000 million (WHO, 2000). The foregoing had actually informed the noble aims and objectives of the following initiatives of the United Nations system: the Global Malaria Control Strategy of 1992; the malaria prevention and control "Action Plan" of 1995; the Harare Declaration on malaria prevention and control of 1997 and the Multilateral Initiative on Malaria in Africa (MIM) of 1998. Again in 1998, the newly elected Director General of WHO initiated the impressive Roll Back Malaria Project, thus establishing malaria as one of WHO's highest priorities.

Onchocerciasis (River-blindness)

Onchocerciasis is a parasitic disease caused by a parasitic filarial nematode or worm, *Onchocerca volvulus*. The disease is transmitted by the bite of a female black-fly of the family simuliidae, specifically *Simulium damnosum* which breeds in fast-flowing rivers, hence the common name river blindness (Ngoumou and Walsh, 1993). In nature, humans are the only known reservoir of this disease. Adult worms, found in the tissue nodules, produce millions of microfilariae that migrate throughout the body, concentrating particularly in the skin and eye. The adult worms may live up to 19 years in the human host (Taylor et al, 1990). The disease is found in 27 countries of Africa where over 99% of all infected persons (mostly farmers) live, and in isolated foci in Latin America and in Yemen in the Arabian Peninsula (Ngoumou and Walsh, 1993). The geographical distribution of the disease in Africa is strictly dependent on the distribution of the man-biting vector, which lies approximately within the area bounded by the two 15 degrees parallels of latitude (Fig. 1b), with a northern extremity at the Abu Hamed, in Sudan, at 19 degrees north and a southern outpost at 17 degrees south in the Cholo district of Malawi (Watson, 1979).

It is estimated that in the endemic areas, over 80 million people are at risk of infection, some 18 million infected and one million people visually impaired, of whom 340,000 are blind (WHO, 1987). An unknown but large number of people also suffer from river blindness-associated severe skin disease, maddening itching which cause great suffering to millions of people (WHO, 1997). Onchocerciasis is a socially and economically destructive tropical disease, which destroys human dignity. This "scourge of the River" also causes debilitation and misery and fear. It kills villages and disrupts family life. According to Watson (1979), the disease not only makes socio-economic development of the largely farming rural areas of sub-Saharan Africa impossible; it also threatens human survival in the worst areas of its endemicity. In hyper-endemic areas, more than 40% of the inhabitants above the age of 40 are blind, and this high rate of blindness has a devastating socio-economic impact. More than half of the inhabitants in hyper-endemic areas will become blind before death, and life expectancy of those who are blind is one-third that of their peers with healthy sights (Taylor, 1990). The frightened population, fleeing their rich but lethal valleys, crowd into and overpopulate the safe plateau lands between the rivers where soils are poor and unproductive. For these agricultural and cattle-rearing folk, and for their governments, it is a disaster.

Dracunculiasis (Guinea Worm Disease)

Dracunculiasis or Dracontiasis, or quite simply Guinea Worm Disease is a crippling water-associated disease, which causes guinea worm infection in man and consequently considerable adverse socio-economic and public health problems on virtually all aspects of human life. The disease, which is caused by Dracunculus medinensis, is one of the largest sub-cutaneous tissue-dwelling parasitic nematodes. Dracunculiasis constitutes a major public health problem in Africa, especially in West Africa (Fig. 1c), with over 15 million people now infected. Dracunculiasis has been reported (Belcher et al, 1975; Nwosu et al, 1982) to enervate large populations, leading to low agricultural productivity, disability and absenteeism of school children and civil servants from school and work respectively. In a report on the economic implications of the guinea worm devastation (Nwosu, 1989) indicated that in an agricultural community of 1.6 million persons in a part of southern Nigeria, the rice farmers alone lost about $ 20 million per annum on account of incapacitation by guinea worm disease at harvest time.

Dracunculiasis is usually found in the remote rural areas where stagnant surface waters such as streams and ponds are the exclusive sources of drinking water. Infection occurs when the drinking water containing Cyclops infected with the third-stage larvae of Dracunculus medinensis is ingested. Although the disease is not directly fatal, it has serious public health implications (Muller, 1971). The infection can incapacitate a sufferer for up to three months. The peak period of potency often coincides with the peak of agricultural activities (Edungbola et al, 1986). The first signs of the disease appear when the gravid female is about to emerge from the subcutaneous tissues. It then provokes the formation of a small blister accompanied by itching and intense burning pain, together with allergic reactions in the form of generalised urticaris, fever giddiness, gastro-intestinal symptoms and dyspnoea. Some of the major complications of the disease include ulcers prone to tetanus infections (Lauckner et al, 1961) and habitual abortion in pregnant women (St. George, 1975).

During periods of heavy physical exertion in the farms in February to April and in June to September, the thirst-inducing activities (land preparation, planting, weeding and harvesting) result in the consumption of large quantities of water from ponds near the farms by the farmers (Belcher et al, 1975). This accounts for the greater exposure and hence prevalence of the disease among the farmers and older children (10-14) who assist them in the farms, as observed by Kale (1977). The habit of farmers drinking directly from ponds and filling their gourds en route to the farms sites complicates the whole process of prevention and control, because it makes no sense of the boiling and filtering advice in the homes.

Schistosomiasis (Bilharziasis)

Schistosomiasis caused by trematodes of the genus schistosoma constitutes one of the important parasitic diseases of man. It affects about 200million people in 74 developing countries, and between 500-600 million others are exposed to infection because of poverty, ignorance, poor housing, substandard hygienic practices and few, if any, sanitary facilities (WHO 1984). In Africa, the planorbid snail transmits S. *haematobium.* Bulinus species including B. *globosus, B. africanus, B. nasatus and B. truncates, B. forskalli and B. senegalensis* have also been suspected to be intermediate hosts of S. haematubium (Betterton et al, 1983; Ukoli, 1990). The endemicity of schistosomiasis in several regions of sub-Saharan Africa and the association between the occurrence of the disease and development of irrigated agriculture and other types of water resources development has also been reported (Audibert et al, 1990; Cowper, 1973).

Schistosomiasis is a blood fluke (nematode) infection with adult male and female worms living in mesenteric or vesical veins of the host over a life span of many years. Eggs produce minute granulomata and scars in organs where they lodge or are deposited. Symptomatology is related to the number and location of the parasites in the human host; Schistoma masoni and Schistoma japonicum give rise primarily to hepatic and intestinal symptoms, including diarrhoea, abdominal pain and hepatosplenomegally and Schistoma haematobium to urinary manifestation, including dysuria, and terminal haematuria. The most important pathological effects are the complications that arise from chronic infection, liver fibrosis and portal hypertension in the intestinal form while obstructive uropathy, superimposed bacterial infection and possibly bladder cancer occur in the urinary form of schistosomiasis. Infection is acquired from water containing free-swimming larval forms (cercariae), which have developed in snails. The eggs hatch in water and the liberated larvae (miricidia) penetrate into suitable freshwater snail hosts. After several weeks, usually while the farmer is working, swimming or wading in water, the miricidia enter the bloodstream, are carried to the blood vessels of the lungs, migrate to the liver, develop to maturity, and then migrate to veins of the abdominal cavity.

Lymphatic filariasis

Lymphatic filariasis (commonly called Elephantiasis) is an important public health and socio-economic disease affecting over 120 million people world-wide. Africa, particularly sub-Saharan region, accounts for nearly 50% of the global burden of filariasis and India for another 35% (Anon, 1998). Anopheles mosquito is the known transmitting agent of the worms, which cause the disease. Filariasis has important social and economic consequences, such as severe stigma associated with lymphoedema and severe reactive skin lesions, and significant reduction in the productivity of affected persons (usually and mostly farmers). The disease prevents farmers from doing their normal farming activities and makes them suffer severe pain, and marriage problems. A study carried out in the year 2000 to assess the extent of "morbidity due to lymphatic filariasis in Northern Nigeria" (Ameh, 2002) showed that out of the 353 patients registered in Zing local government area of Taraba state, 68% were females while the remaining one-third were males. "Women with the problem are often divorced. This worsens their economic ability and leads to psycho-social pains at home and within the community" as they are often teased and called "alede" (meaning pig).

Diarrheal diseases

Diarrhea is a clinical syndrome of diverse etiology, associated with loose, or watery stools and often vomiting and fever. It is a symptom of bacterial, viral and parasitic enteric agents causing cholera, shigellosis, salmonellosis, yersiniosis, giardiasis and viral gastroenteropathy. It can be caused by infection with *Escherchia coli* and campylobacter strains, non-cholera vibrious and vibrio parachaemoyticus, other infectious diseases such as malaria measles, and intestinal helminthes as well as chemical agents. Out-breaks usually occur in areas with poor sanitation, and with common sources of food and water. Major mode of transmission is faecal contamination of food water or formities in the area. Cholera is an acute bacterial enteric disease with sudden onset, profuse watery stools, occasional vomiting, rapid dehydration, acidosis and circulatory collapse. This is caused by vibrio cholerae, which is transmitted primarily through ingestion of water contaminated with faeces or vomitus of patients. Raw and under-cooked foods from polluted waters cause outbreaks or epidemics in rural sub-Saharan Africa.

Giardiasis is a protozoan infection principally of the upper small intestine; while often asymptomatic, it may also be associated with a variety of intestinal symptoms such as chronic diarrhea, steatorrhea, abdominal cramps, bloating, frequent loose and pale greasy stools, fatigue and weight loss. Mal-absorption of fats and of fat-soluble vitamins may occur. Damage to duodenal and jejunal mucosa cells may occur in severe giardiasis. Localised outbreaks occur from ingestion of cysts in fecally contaminated water and less often from fecally contaminated food.

Respiratory Diseases (Influenza and Tuberculosis)

Influenza is an acute viral disease of the respiratory tract characterised by fever, chilliness, headache, mylgia, prostration, coryza and mild sore throat. Cough is often severe and protracted and usually self-limited with recovery in 2-7 days. The common cold, cough and viral pneumonia and undifferentiated acute respiratory diseases are often caused by influenza. Gastrointestinal tract manifestations (nausea, vomiting, diarrhea) are rare, but have been reported in up to 25% of children (one form of farm labour) in school outbreaks of influenza. Influenza derives its importance from the rapidity with which epidemics evolve, the widespread morbidity, and the seriousness of complications, notably viral and bacterial pneumonias. During major epidemics, severe disease and deaths occur primarily among the elderly and debilitated by chronic cardiac, pulmonary, renal or metabolic disorders.

Tuberculosis is a mycobacterial disease important as a cause of disability and death in many parts of the world. The initial infection usually goes unnoticed; tuberculin sensitivity appears within a few weeks; lesions commonly heal leaving no residual changes except pulmonary or tracho-bronchial lymph node calcifications. It may progress directly to pulmonary tuberculosis or, by lymphohematogenous dissemination of bacilli, to produce pulmonary, milliary, and meningeal or other extra pulmonary involvement. Serious outcome of the initial infection is more frequent in infants, adolescents and young adults.

The challenge the effects and the driving forces

The problem of poor health and poor agricultural productivity is a challenging one. It leaves in its tail a sting of a dilemma. On the whole, it can be seen that producing our food is often bad for our health, just as poor health hampers us in growing our food. And recovering from high levels of fatalities, disabilities and injuries in agricultural food production and defeating such diseases as malaria, onchocerciasis, schistosomiasis, tuberculosis and the killer HIV/AIDS, etc. can clash with agricultural practice. Premature death and spells of illness from infectious diseases cuts down tomorrow's and today's labour force. It is no wonder then, that the re-emergence of certain diseases brings not only shudder to the body of politics, and the body of economics of a nation.

The overall effect of these and other communicable diseases, such as dengue fever, or the yellow fever recently established in some parts of eastern Africa, is not only a tragedy of lives lost early, but also the debilitating effect on a nation's health and welfare of a debilitated youth and a debilitated labour force. Since communicable diseases are rife in developing countries (developed ones are more prone to death from lifestyle illnesses, such as cancer and cardiovascular diseases), it is agriculture, which suffers the most. In the field, or the workshop, the energy and concentration of a person is sapped by illness. Accidents are more frequent. A weakened body is further worn down by the strain of ill-designed equipment and the misuse of chemicals, pesticides and processing. Some diseases, such as malaria, are often most prevalent at harvest time, and play havoc with the work force.

The idyllic picture that some urbanites have of rural life is not the same one that we see through the mist of pain and sweat, and our exposure to physical, chemical and, above all, biological hazards. In the fields, the harvest store and the fish pond, we stand unprotected from the many dangers, literally bugs and bacteria, that form what the experts call "micro-biological production systems" – breeding grounds, in short. Of such irresolvable clashes are made the endless agonies of the rural poor, for they will suffer one way or the other as they always do (they already spend 40% of scarce disposable income on malaria 'cures'), and less pitifully, the headaches of the policy-makers. It seems that the only real policy in achieving co-existence between the two sectors is to apply the art of compromise, not always the most obvious skill of notice to policy-makers.

It is necessary to understand the forces which are driving the diseases epidemic, or rather pandemic, now that is global. Some people present HIV/AIDS as a disease in terms of microbes; many people associate its spread with commercial sex workers and loose sexual behaviour. It is not that simple. There are also considerations of family and survival. In so many of the communities HIV/AIDS is not about the virus, it is about food. In several communities, rains come only three months every year. By April, the place is dry and the harvests are in the store, but by end of August, the food stores have run out and you have families without food. So if a father has a teenage girl and she goes out and comes back with a kilogram of rice the father hardly will ask "why". It is also a question of risk. People living in comfort perceive risk in terms of the future, tomorrow, the year after ten years, twenty years; it is theoretical because you know that you have something to live for. But talk about HIV/AIDS to a commercial sex worker, driven to her work by desperation, and she will say that hunger tonight is a greater risk to her and her children than HIV/AIDS, perhaps, after five years.

The concept of less poverty, less disease is apt and applicable here. As Louis Pasteur said, 'The microbe is nothing, the terrain is everything'. That Africa is the poorest continent and that Africa is facing the biggest epidemic of HIV/AIDS and other diseases surely tells us something about the relationship between poverty and the disease. The question now is; how much further will the diseases dig into an agricultural sector that is already prey to under-nourishment in terms of income and input and to the ill health of its people. Instead of condemning the alleged promiscuity of our youth, who say their behaviour is only a symptom of poverty and absence of development, we need to commence action(s) aimed at ensuring that disease prevention and poverty alleviation go hand-in-hand.

Conclusions

Sustainable development of agriculture and the management of resultant health conditions in the sub-Saharan Africa will depend mainly in changes of behaviours and attitudes on the part of the people. The same applies to coping with the permanent impact of disease conditions on agriculture and rural development in sub-Saharan Africa. To make the necessary changes in agricultural practices requires changes in attitudes in the individuals who make up its many institutions. Informal institutions in the rural areas will have to change, and the process of change will have to the encouraged by local, regional, and national governments, the private sector and donor agencies. These will need to re-invent forgotten social mechanisms of being non-bureaucratic and supporting small rural projects. For formal bodies, understanding the changed needs and nature of the farmer means listening, asking and understanding at local level through more participatory approaches.

Seeking and making available solutions, such as adapting technologies for solving farmers' problems and providing regulatory climate governing the release and use of new biotechnology products means establishing genuine partnerships between institutions (for example, agriculture and the judiciary). In many countries, the existing legislation is adequate to regulate the use of most agricultural products, which are likely to be produced using bio-technology. The new requirement is for guidelines to cover the handling of genetically engineered organisms (GMOs) or genetically modified products (GMs) at the experimental stage, and for the assessment of any risk associated with the commercial release of new organisms into the environment. The guidelines should be framed so as to complement and support those endorsed by existing regulatory agencies such as the Organisation for Economic Co-operation and Development (OECD) and International Labour Organisation (ILO). The poor should be included directly in the debate and decision making about their desire for technological change, the risks of that change, and the consequences of no, or alternative, kinds of change.

The disease prevention and control strategies stress the selective use of preventive measures wherever they can lead to sustainable results. Such measures should be aimed at halting the deterioration of the disease situation, contributing appropriately to the development of health services, and encouraging inter-vector (agriculture and health) co-operation and community participation. Implementation of control measures depends on a number of factors, such as disease epidemiology, available information on clinical manifestations of disease conditions, the potentials and limitations of each method of control, infrastructure and institutional capacities. These vary within, and among, countries and regions. In all, the major strategies should aim at (a) Surveillance and information dissemination to farmers and their families on the appropriate and sustainable agricultural development practices, (b) A decentralised program of management for prevention and control of common diseases associated with such practices, (c) Care and support for people living with all debilitating diseases, including HIV/AIDS, malaria and tuberculosis and (d) Providing blood safety and laboratory support for rural farmers and community members.

Surprisingly, most diseases control programs in Africa south of the Sahara have a limited capability for planning disease control activities. They also have limited human and financial resources to execute even well planned disease control activities. Even when these countries have national biomedical research organisations, there is often little communication or collaboration between these and the control programs. While sound, safe and sustainable agricultural development practices remain the watchwords, the policy considerations for disease prevention and control should include (a) Strengthening early diagnosis and prompt treatment of diseases to reduce morbidity and mortality in high risk groups, especially farmers and improving the quality of care in the home, (b) Improving disease management in the resource-poor farmers' communities, small-scale, medium and large-scale farming establishments, (c) Strengthening the capacity for implementing selective preventive measures and reducing the impact of ill-health on farmers generally, especially pregnant women and children (d) Increasing the use of effective personal protection measures by farmers and their extended family members, (e) Strengthening community involvement and participatory approach in disease control; using especially opinion and youth leaders who should be trained on the advantages of disease prevention at rural/village community levels and (f) Strengthening national capacity to predict, detect early incidences and prevent epidemics.

In the implementation of the above measures, three important social factors are necessary; involvement and active participation of all the stakeholders (agricultural research scientists, bio-technologists, medical, biomedical and public health officers, and the farmers), proper understanding of the local perspectives, indigenous health-care technical knowledge and practices, and proper understanding of the peculiarities of the farming communities' socio-economic and biophysical environments. Partnerships, networking, and the promotion of equal opportunities in endemic communities will always constitute the core of disease control and prevention strategies. All the stakeholders, again working in collaboration, should lead in concert in the analysis of past and current agricultural development programs in the sub-Saharan Africa and their impacts on the health of citizens. This should enable experts to identify causes of health problems associated with agricultural development and new products in the region and develop appropriate interventions to reverse or modify underlying factors and define or re-define future actions.

Major diseases in Africa

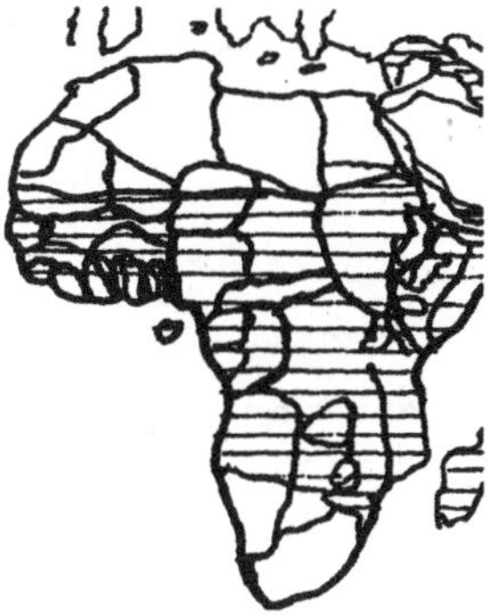

Fig. 1a: Areas in Africa where malaria transmission occurs.
Source: WHO (2000)

Fig. 1b: The shaded area of the map shows the onchocerciasis-region of Africa. Source: Ngoumou and Walsh (1993)

Fig. 1c: Map showing the distribution of dracunculus medinensis in Africa.
Source: Ukoli (1990)

Table 1: Possible impacts of HIV/AIDS on agricultural-dependent households

- □ Adult becomes sick.
- □ She/he reduces work.
- □ Replacement labour is "imported," perhaps from relatives.
- □ All adults work longer hours on the farm.
- □ Healthcare expenses rises (e.g., drugs, transport).
- □ Household reduces food consumption.
- □ Household switches to less labour-intensive crops and farming systems, small livestock.
- □ Nutritional status of sick adult deteriorates.
- □ Sick adult stops work.
- □ Family members spend more time caring for the sick adult, less time on child care.
- □ Divisible assets (e.g. livestock) are sold.
- □ Debts increase.
- □ Children drop out of school to help with household labour.
- □ Sick adult dies.
- □ Household incurs funeral expenses.
- □ Household may fragment as other adults migrate for work.
- □ Household reduces cultivation of land; more land is left fallow.
- □ Inappropriate natural resources management may lead to increased spread of pests and diseases.
- □ Effects of loss of farming knowledge intensify.
- □ Mining of common property resources increases.
- □ Access to household land and property (particularly for surviving widows) may be affected.
- □ Solidarity networks are strained, possibly to the point of collapse.
- □ Surviving partner becomes sick.
- □ The downward spiral accelerates.

Source: Gillespie et al (2001).

Table 2: Regional HIV/AIDS statistics and features (at the end of 2000) and atures

Region	Epidemic started during	Adults & children living with HIV/AIDS disease	Adults & children newly in-fected with HIV	Adults prevalence rate % (*)	Percentage of HIV posi-tive adults who are women	Main mode(s) of transmission for adults living with HIV/AIDS
Sub-Saharan Africa	late 70s, early 80s	25.3 million	3.8 million	8.8 %	55 %	hetero
North Africa & Middle East	late 80s	400,000	80,000	0.2 %	40 %	Hetero, IDU
South East & South Asia	late 80s	5.8 million	780,000	0.56 %	35 %	Hetero, IDU
East Asia & Pacific	late 80s	640,000	130,000	0.07 %	13 %	IDU, Hetero, MSM
Latin America	late 70s, early 80s	1.4 million	150,000	0.50 %	25 %	MSM, IDU, Hetero
Caribbean	late 70s, early 80s	390,000	60,000	2.3 %	35 %	Hetero, MSM
Eastern Europe & Central Asia	early 90s	700,000	250,000	0.35 %	25 %	IDU
Western Europe	late 70s, early 80s	540,000	30,000	0.24 %	25 %	MSM, IDU
North America	late 70s, early 80s	920,000	45,000	0.6 %	20 %	MSM, IDU, Hetero
Australia and New Zealand	late 70s, early 80s	15,000	5,000	0.13 %	10 %	MSM
Total		36.1 million	5.3 million	1.1 %	47 %	

* The proportion of adults (15 to 49 years of age) living with HIV/AIDS in 2000, using 2000 population numbers. Hetero (hetero sexual transmission), IDU (transmission through injecting drug use), MSM (sexual transmission among men who have sex with men). Source: Anon (2001), HIV/AIDS: Global crisis, Global action. Cover story. In: Partnership For Development, Vol. 2, No. 1, December 2001.

References

Ameh, F. (2002). "Don Calls for Support for Elephantiasis' Victims". *Punch Newspaper,* 17 (18): 634. Monday, August 26, p.12.

Anon, (1996). "Antimalarials," pp. 457-477. Martindale: The Extra Pharmacopoeia, 31st Edition. J. E. F. Reynolds (Editor), Royal Pharmaceutical Society, London.

Anon, (1998). "Community-directed Treatment of Lymphatic Filariasis." Africa. Draft Report of a Multi-centre Study in Ghana and Kenya. UNDP/World Bank/WHO - TDR/DIE/FIL/00.1.

Anon, (2001). "HIV/AIDS: Global Crisis, Global Action. Cover Story." Partnership for Development, 2(1): 4. December 2001.

Audibert, M., R. Josseran, and A. Adjidji. 1990. "Irrigation, Schistosomiasis and Malaria in the Lagone Valley, Cameroon." Amer. Jour of Trop. Med. and Hyg. 42(6): 550-560.

Barnett, T. and G. Rugalema. 2001. "HIV/AIDS: A Critical Health and Development Issue." pp. 43-47. The Unfinished Agenda; Perspectives on Overcoming Hunger, Poverty, and Environmental Degradation. Per Pinstrup-Andersen and Rajul Pandya-Lorch (eds.), International Food Policy Research Institute, Washington, DC. 301pp.

Belcher, D. W., F. R. Wurapa, W. B. Ward, and I. E. Lourie.1975. "Guinea Worm in Southern Ghana: Its Epidemiology and Impact on Agricultural Poductivity." Amer. Jour. Trop. Med. and Hyg. 24; 243-249.

Benenson, A. S. 1985. *Control of Communicable Diseases in Man.* American Public Health.

Betterton, C., S. E. Fryer and, C. A. Wright. 1983. "Bulinus senegalensis (Molusca; Planorbiae) in Northern Nigeria." *Ann. Trop. Parasitol.* 77(2): 143-149.

Bongaats, J. and J. Bruce. 2001. "Population growth and policy options." *The Developing World,* pp. 61-66. In: *The Unfinished Agenda: Perspectives on Overcoming Hunger, Poverty, and Environmental Degradation.* Per Pinstrup-Andersen and Rajul Pandya-Lorch (eds). International Food Policy Research Institute, Washington, DC., 301 pp.

Cowper, S. G. 1973. "Bilharziasis (schistosomiasis) in Nigeria.". 25,105-118.

Edungbola, L. D., S. J. Watts, O. O. Kale, C. S. Smith, and H. R. Hopkins. 1986. "*A Method of Rapid Assessment of the Distribution and Endemicity of* Dracunculiasis in Nigeria." *Soc. Sci. Med.* 22(6): 555-558.

Fleshman, M. 2001. "AIDS Orphans: Facing Africa's 'Silent Crisis'". *Africa Recovery.* 15(3):1-20: A Publication of the United Nations Department of Public Information.

Gillespie, S., L. Haddad, and R. Jackson. 2001. "HIV/AIDS, Food and Nutrition Security: Impacts and Actions." *Nutrition and HIV/AIDS, Nutrition Policy Paper*, No. 20. ACC/SCN: Geneva. International Institute of Tropical Agriculture (1993). "*Improving Life in the Humid African Tropics: Eco-regional Programme for the Humid and Sub-humid Tropics of sub-Saharan Africa*. IITA, Ibadan. 17 pp.

Kale, O. O. 1977. "The Clinico-epidemiological Profile of Guinea Worm in the Ibadan District, *Nigeria." Amer. Jour. Trop. Med. and Hyg.* 26: 208-214.

Lauckner, T. L., A. M. Rankin and, F. C. Ali. 1961. "Analysis of Medical Admissions to the University College Hospital, Ibadan." *West African Medical Journal* 10: 3.

Minc, L. D. and J. Vandermeer. 1990. "The Origin and Spread of Agriculture." *Agro Ecology*. C. R. Carrol, J. H. Vandermeer, and P. M. Rosset (eds.). McGraw Hill Publication Co., pp. 65-111.

Mills, C. F. 1987. "The Detection of Trace Element Problems in Developing Countries." *Elements in Health and Diseases*. M. Said, M. A. Rahrnan, and L. A. D'Silva (eds.), pp. 74-89, Hamdard University Press, Karachi.

Muller, R. 1971. "Dracunculus and Dracunculiasis", *Advances in Para-sitology* 9:83-151.

Ninja, F. C. 2001. "HIV/AIDS: Global Crisis, Global Action. Cover story." *Partnership for Development* 2(1): 4-7. Dec. 2001.

Ngoumou, P. and J. F. Walsh. 1993. *A Manual for Rapid Epidemiology Mapping of Onchocerciasis*. UNDP/World Bank/WHO-TDR/ TDE/ ONCHO/93.4. 37 pp.

Nwosu, A.B.C. 1989. "Economic Aspects of the Connection Between Agricultural Production and Guinea Worm Infection." *Nig. Jour. Parasitol* 10 (Suppl. l): 24-33.

Nwosu, A. B. C., E. O. Ifezulike, and A. O. Anya. 1982. "Endemic Dracontiasis in Anambra State in Nigeria; Geographical Distribution, Clinical Features, Epidemiology and Socio-economic Impacts of the Disease. *Ann. Trop. Med. Parasitol* 72, 187-200.

Spore. 1999a. "Urban Agriculture and Food Security: Take Your Farm to Town." *Information for Agricultural Development in ACP Countries* 81, 1-2. June 1999.

—— 1999b. "Aids and Agriculture: Red-ribbon Farming." *Information for Agricultural Development in ACP Countries*. 82, 1-2. August 1999.

——2000. "The AIDS Pandemic and Agriculture: The Microbe and the Macro Policy. *Information for Agricultural Development in ACP Countries* 89, 16. October 2000.

——2001. "Occupational Health and Safety in Agriculture: Safety Can be the Hardest Word. *Information for Agricultural Development in ACP Countries* 91, 1-2. February 2001.

—— 2002. "Poor Health, Poor Agriculture." *Information for Agricultural Development in ACP Countries* 97, 1-2. February 2002.

St. George, J. 1975. "Bleeding in Pregnancy due to Retro Placental Situation of Guinea Worm." *Annals of Trop. Med. and Parasitol* 69: 383-386.

Taylor, H. R. 1990. "Onchocerciasis." *International Ophthalmology* 14: 189-194.

Taylor, H. R., M. Pacque, B. Munoz, and B. M. Greene. 1990. "Impact of Mass Treatment of Onchocerciasis with Ivermectin on the Transmission of Infection." *Science* 250, 116-126.

Ukoli, F. M. A. 1990. *Introduction to Parasitology in Tropical Africa.* Textflow Ltd., Ibadan, 464 pp.

Watson, J. 1979. "Onchocerciasis; Scourge of the River." *Africa Health* pp. 31-34. February 1979.

Wilson, E. 2001. "The Population Boom: What do the Numbers Mean?"pp 57-60. *The Unfinished Agenda: Perspectives on Overcoming Hunger, Poverty, and Environmental Degradation.* Per Pinstrup-Andersen and Rajul Pandya-Lorch (eds.). International Food Policy Research Institute, Washington, D.C., 301 pp.

World Health Organisation. 1984. "Global distribution of schistosomiasis." GEGET/WHO Atlas. *World Health Statistics Quarterly*, 37: 186-199.

———1987. "World Health Organisation Expert Committee on Onchocerciasis." *Third Report, Technical Report Series*, 752, WHO, Geneva.

———1997. "Onchocerciasis." *Prospects for the Elimination of some TDR Diseases*. pp 11-16; TDR/GEN/97.1. 35pp.

———2002. WHO Expert Committee on Malaria: Twentieth Report. *WHO Technical Report Series 892.* Geneva 2000. 75pp.

LINKING POVERTY TO FOOD INSECURITY AND MALNUTRITION IN SUB-SAHARAN AFRICA

Susan Keino

Abstract

Poverty, food insecurity and malnutrition are among the challenges of the 21st century in developing countries. In the wake of the 21st century Africa has faced many challenges including declining economies, high population growth, disease outbreak, the HIV/AIDS pandemic, the outbreaks of famines caused by either natural disasters (e.g. droughts) human activities (e.g. civil wars), which have resulted in breakdowns of food production and distribution systems. All these factors among others have conspired to make Africa the most food insecure and poor region with increased cases of malnourished individuals. Although access to and availability of food has been declared a human right in sub-Saharan Africa, this is far from becoming a reality. This right to food imposes obligations on states to respect, protect, fulfil and promote food security, however, in Africa most governments are facing other challenges such as conflicts with other nations, civil wars etc., such that meeting the food needs of its people is no longer a priority. However, as much as poverty reduction programmes have been in place for some time now, their efforts have not gone hand in hand with improved nutritional status, although it has been well documented that poverty is the principle cause of malnutrition.

This paper, therefore, will discuss the intricate relationship between poverty, food insecurity and malnutrition, compare the intervention strategies as well as introduce agriculture as the main component overlooked in these strategies. It will also look at emerging issues in the concept of food security in sub-Saharan Africa.

Introduction

According to FAO statistics (1997) current estimates indicate that approximately 40% of Africa's population, are chronically undernourished. This translates into over 200 million people. Many factors have conspired to reinforce the vicious circle of poverty, food insecurity and malnutrition in the region. These factors include increase in poverty, persistent droughts, weak governance, wars and civil conflicts, high population growth, prevalence of disease and the HIV/AIDS pandemic, inadequate technology (seen in poor farming methods and post-harvest handling of farm produce etc.), inefficient transport infrastructure, discriminatory trade terms as well as unsustainable management and utilisation of natural resources among many other factors.

The World Food Summit held in Rome in November 1996 recognized the fact that access to and availability of food is a human right and the right to food imposes obligations on states to respect, protect, fulfil and promote food security (Maxwell and Frankenberger, 1992). However, the right to food is an example of the most commonly expressed right as well as the most commonly violated one (ACC/SCN, 1997). The same summit acknowledged that in Africa the solution to problems of poverty and malnutrition lies in the empowering of rural communities by assisting them in identifying and overcoming constraints to participating in developmental processes that will eventually lead to improved living standards.

There have been diverse efforts to eradicate poverty in developing countries. In as much as these efforts have been in place in some countries for the last 20 years, there have been improved living standards in few of these countries except in sub-Saharan Africa where the progress has been slow (ACC/SCN, 2000). Many countries have experienced growth in gross domestic product, food production and improvements in school enrolment, gains in status of women, improved access to primary health care and clean water, and an increased respect for human rights. The same is not happening in sub-Saharan Africa. One of the likely reasons why this is not happening could be of the lack of understanding of the interrelationship between poverty, malnutrition and food insecurity. Secondly, in the developing countries particularly in Africa there is lack of sustained political commitment and a great deal of managerial incompetence. This is obvious considering that most governments today commit a little more than 10% of their budget to directly meeting the basic needs of their people. More money is still being spent on military capacity and on debt servicing than on health and education (UNICEF, 1993).

Ironically, the world's resources put together can meet the needs of almost every one on the planet for adequate food, safe water, primary health care, family planning and basic education, yet scores of people are dying as a result of hunger. It is also sad to realize that meeting the needs of the poorest and the least politically influential nations in the developing world has rarely been a priority of governments particularly in the developed world. These are the same governments that make up the intergovernmental organisations (IGO's), which have the powers to influence decisions made in forums convened to address these life-threatening issues that have adversely affected the developing nations.

The poverty that has plagued sub-Saharan Africa is multidimensional. It has different faces depending on which side you are looking at it. However, what is normally ignored is the fact that poverty in Africa is an unwanted inheritance from the previous generations. A child born into poverty may not realize that they are poor until they move out of their home and look back in shock at where they have come from and how far they have come. What they may realize, however, is that someone has to struggle to put food on the table, and sometimes the children have not been spared from this responsibility and as they grow up they realize that they never have enough food to eat no matter how much they struggle. Poor and hungry people may not care much about the means to an end but the end itself. In other words, they are concerned more in meeting their immediate needs; that is getting something to eat. Therefore, a five-year-development plan of poverty eradication or a 2015 mark of halving the number of malnourished and food insecure people in the world may not sound very comforting to the afflicted. World Bank's poverty reduction strategies programmes (PRSP), for the Highly Indebted Poor Countries (HIPC) have been received with a lot of expectations. But they have not addressed the issue of poverty as a crucial link to hunger and malnutrition. Therefore, the right to food concept should be introduced as suggested in SCN News (2001) in order to achieve its objective. The issue of human right to adequate food must become a central objective in development plans of all countries in order to put an end to hunger and malnutrition; the governments and states ought to be held responsible for violating this right. How can the governments (both regional and international) then be urged to co-operate? Co-operate in such manner as doing away with policies that are detrimental such as trade policies that open up markets to products that can otherwise be produced locally, promoting transparency and good governance, ensure that each person has access to food etc. Yet again, when this right is violated where do the hungry run to for help, is it the WFP? We have been quick to set up tribunals to sort out war crimes; however, no one wants to fight for the rights of the hungry people.

Relationship between poverty, food insecurity and malnutrition

Poverty breeds malnutrition and in turn malnutrition increases poverty, a vicious cycle it is (SCN, 1995). Malnutrition in itself has far reaching consequences and it may be one of the main causes of poverty. Poverty on the other hand is stated as the direct cause of malnutrition and that no nutritional improvement can be expected unless and until poverty is reduced (ACC/SCN, 1997). This interrelationship is further illustrated in the figure 1. Whereby malnutrition is seen to result from effects of inability to access food to inadequate food supply as well as inadequate intake of food. The inaccessibility of food and inadequate food supply are known dimensions of food insecurity. The reason for this could be poor infrastructure, low production due to poor farming methods, poor market strategies, political instability etc. These factors in most cases are considered as indicators of poverty. The resulting effects of poverty may in one way or another lead to malnutrition, such is the interrelationship among the three factors. To be able to understand this link one would need to look further into the cause and effect relationship of malnutrition, food insecurity and poverty.

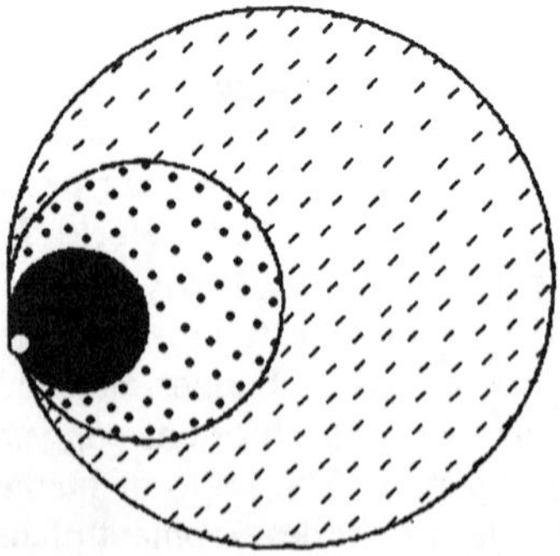

KEY

o common resource base malnutrition

▪ malnutrition

∴ food insecurity poverty

/// poverty

Figur I: Conceptiualising the relationship in poverty alleviation strategies, food insecurity and malnutrition intervention strategies

The far-reaching consequences of malnutrition are evident in our day-to-day lives. Consequently, at least one-fifth of the worldwide loss of years of life to death and to disability is due to undernutrition (ACC/SCN, 2002). In summary undernutrition contributes to poverty directly or indirectly as follows:

- Maternal malnutrition causes intrauterine growth retardation (IUGR), due to this the foetus has a much higher risk of dying in infancy. ACC/SCN (2002) reports that some 30 million infants are born each year in developing countries with impaired growth due to poor nutrition during fetal life.
- Fetal and infant undernutrition affects children's later school enrolment, educational attainment, cognitive ability and lifetime earnings.
- Survivors in this case of IUGR are unlikely to catch up significantly on this lost growth and more likely to experience developmental deficits. Underweight children tend to have severe illnesses, e.g. diarrhea and pneumonia. Their immune system is also compromised by undernutrition and a child will grow up susceptible to infections.
- The consequences of being born malnourished extend into adulthood in that there is a link between maternal and early childhood undernutrition and increased adult risk of various chronic diseases. These diseases have long-term effects and the complications make them expensive to manage.

- Low adults' nutrition status affects their labour productivity, thus, improved adult nutrition will lead to improved health, cut down on medical costs and time lost due to illness or inability to work due to hunger pangs.
- Hunger can be the source of conflict in a community and this could pose a negative impact on economic and political stability as well as on the sustainability of the food systems overall.

The intergenerational cycle of poverty is more likely to be broken when babies get an adequate nutritional head start. Thus, there is the need for improved maternal nutrition as well as nutritional status of children under five years old. Good nutrition is, therefore, argued to be an essential building block for development in that by eliminating malnutrition in children and women, one can be sure of having a population with improved cognitive skills that will generate new resources and technologies and will imagine, create and nurture new institutions and partnerships to improve global welfare. Thus, improved nutrition is central to improved income generation, poverty reduction, and more rapid development.

Nutrition intervention and poverty reduction

While better nutrition empowers people, the process through which malnutrition is reduced can also empower communities. Unlike most other types of human capacity investment interventions, most nutrition programs are community based. As communities develop the capacity to undertake assessment, analysis and action for nutrition interventions (Triple A cycle), they can apply this capacity to other types of development intervention, for example, this can be applied to food insecurity interventions and poverty alleviation programs at the community level for viability and sustainability.

In comparing South Asia and sub-Saharan Africa, South Asia is the poorest region in the world with the highest proportion and number of the world's malnourished children and women. In sub-Saharan Africa 40% of the population is reported to be malnourished and the nutrition situation is not improving. In his paper on Poverty and Nutrition in South Asia (ACC/SCN 1997), Osmani Siddiq indicates that while there are proportionately more poor people in south Asia, they are on the average less poor than their counterparts in the sub-Saharan Africa. The reason is that South Asia has made considerable progress in reducing the incidence of poverty over the years, while sub-Saharan Africa and most of Latin America have stagnated, especially in the last decade. Looking at South Asia closely we find that many countries actively pursued poverty alleviation policies based on programs of asset distribution, empowerment generation and the provision of safety nets.

However, regardless of this, child malnutrition prevalence remains high. The countries that have notably recorded a reduction in incidence of poverty include India and Pakistan have been attributed to distribution of income. Income per se is necessary but not sufficient in itself in improving nutrition. This is because of the following factors: high income may not reduce caloric deficiency in people's diet; the role of women in shaping the nutritional status of household members. Not much can be done if they are unable to make good use of private and public resources. This could be as a result of illiteracy, socio cultural factors such as lack of access to credit facilities, inability to own land, etc. Income can be allocated to non-food items.

The poverty alleviation programs in South Asia are nationally co-ordinated and although significant progress has been made in reduction in poverty, the same have not been reflected in reduction in malnutrition. Hunger has continued to persist among poorest households. Therefore, to be able to understand why such poverty alleviation strategies have not yielded much fruit as far as nutrition is concerned. There is the need to understand the crucial link between poverty, malnutrition, and food insecurity.

However, it is often in the implementation of food security and nutrition interventions that the conceptual links between agriculture, food and nutrition intervention break down. Food security is too often formulated as food production, and little attention is given to determining whether agricultural activities contribute to alleviating hunger and improving overall nutrition status. As well as, nutrition is too often seen as a health issue focusing on biomedical interventions that emphasise supplementation and fortification over food based strategies linked to agricultural production.

In the conceptual model for poverty, food insecurity and malnutrition above (Figure 1), the inner circle represents the presence of malnutrition in a community, which calls for community-based agricultural foods and nutrition intervention strategies. These strategies may in one way or another be linked to the ongoing poverty alleviation strategies. These strategies aim at eliminating food insecurity. Fighting poverty and achieving food security are two sides of the same struggle. Naturally, it is easier for one to move from the inner to the outer circle, than the other way round. The circumference in the outer circle is bigger in comparison to the inner circle, this means that one needs many organisations and many projects which in the long run will all be aiming at improving the welfare of individuals, whereas in the inner circle one needs few organisations and strategies which will translate to use of few resources and will ensure cost effectiveness. All the three circles meet at one point indicating that all efforts be it to alleviate poverty, food insecurity or malnutrition have to depend on one resource base. In a nutshell, there is need for the implementation of many projects at the community level, handling malnutrition issues rather than poverty alleviation projects at the national level, as it is the case in many developing countries. This approach (the in-out approach) requires a community that has realized and prioritised its problems. The needs of the community have to be respected by the organisations facilitating the community project. Examples of such projects that have been seen to work are cited in SCN News (No. 15 Dec.1997). These include projects that have taken malnourished individuals, educated them on proper eating habits; production and consumption of foods rich in certain nutrients; hygiene and sanitation; intervention strategies for the observed cases in malnourished individuals such as VAD, IDA and IDD through providing capsules and diet management; and also through health care to handle clinical cases. It should be noted, that these projects would be going on at different levels in different communities, the same community, in one or several countries (e.g. those aimed at improving food production and those of hygiene and sanitation may be going on at the same time and in the same community).

Examples of such projects and the countries in which they have been implemented include:

i) Community food and nutrition programme in Zimbabwe
ii) Nutrition surveillance and intervention in South Africa
iii) Effective programmes for improving nutrition in Ghana
iv) Child survival protection and development programme (CSPD) in Tanzania.

These programmes may seem like a kind of emergency programmes, whose results are expected within a short time and may not seem to be sustainable at a glance. However, the whole essence of the community-based approach is to benefit the community but in the long run it will benefit the whole country. In other words, their gradual and step-by-step process is necessary, given that the major objective is to achieve countrywide coverage to relieve the suffering of the greatest number of people within a short period of time and in a cost effective manner. To make these community projects a success, what was observed as necessary common features in the above projects, was the need for political commitment and support mechanisms to be put in place.

David Beckmann (President, Bread for the World) sums it all up well when he says that nutrition programs allow the reduction of hunger more quickly than poverty alleviation programs in general (SCN NEWS, 2001). Therefore, strong measures have to be put in place in support of nutrition intervention programs, as they are far more important than those aimed at alleviating poverty directly. The illustrated link in the intervention strategies can be observed in the schematic diagram below (Figure 2).

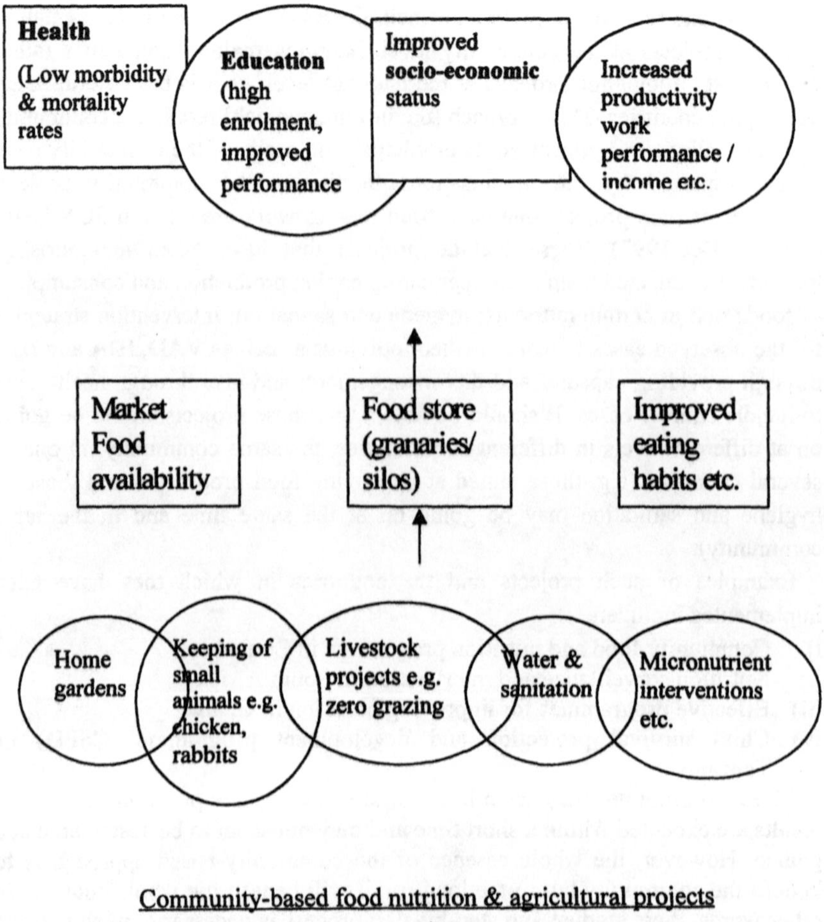

RESOURCE BASE

(Infrastructure, land, capital, indigenous knowledge provided collectively by NGOs, governments, communities, etc.)

Figure 2: Integration of community-based, agricultural, food and nutrition programmes

The concept of food insecurity : a brief look at emerging issues towards a sustainable action

According to Ogunrinade et al (1999), the concept of food security differs fundamentally from food production. Food security per se means ensuring that all people have physical and economic access to the basic food they need to work and function normally. It can be constrained by problems of availability (inadequate supplies of food, including imports of food aid) and by difficulties in physical and economic access. Physical access to food can be impeded by poor infrastructure, poor marketing and storage facilities, and strife. Economic access has two dimensions: the ability of nations to generate the foreign exchange to pay for food imports and the ability of the household to generate the income necessary to buy food. However, famines and food shortages resulting from political strife, war, floods, droughts or poor harvests have threatened food security in sub-Saharan Africa. In addition, in sub-Saharan Africa where agriculture constitutes the livelihood of 69 per cent of the economically active population (FAO, 1997), one would expect improved agricultural activities and technology as well as establishment of agro-processing industries to significantly have an impact on food insecurity in the region but this has not been the case.

Localisation of hunger in Africa is alarming. Sub-Saharan Africa has been by-passed by many of the new forces of globalisation. However, hunger persists in Africa not because of global market malfunctions, but instead because of low productivity growth within Africa's own farming sector. Persistent drought, civil strife, impacts of HIV/AIDS as well as high population growth has made it impossible for Africa to be self-sufficient in food production. FAO (1997) confirms that food and farm production per capita is declining in Africa. Declining agricultural production is the main problem in Africa, because agriculture for many countries is the backbone of the economy such that when it lags it translates directly into lagging rural income growth, persistent poverty and, hence, persistent hunger. In one way or another ruling governments are to blame in many countries. This is because they have failed to provide essential public goods such as internal peace, rule of law, protection for individuals or community property, adequate rural infrastructure and sufficient investment in agricultural research and fair trading policies.

Therefore, it is the governments in Africa that must take primary action in ending hunger. What does it take to end hunger in Africa? So far we have had such strategies as foreign aid, structural adjustments, policy reforms, capacity building etc., but how have we fared in these areas? In as much as these strategies have been in place for quite some time, hunger has continued to persist.

We know that food-based strategies are the key to addressing global hunger and malnutrition. What do we mean exactly by “food-based strategies”? These include biodiversity, mixed farming, intercropping among others. While pursuing these strategies one needs to bear in mind the need to feed a growing population and the use of small and unproductive lands as well as declining labour supply due to increased mortality due to diseases and the HIV/AIDS pandemic. Thus, the need for a sustainable food availability and access even under the prevailing circumstances.

It has been said that most of the food challenges of the growing population can be met through application of science and technology and that technology has been and will always be a powerful tool for human development and poverty reduction. But how true is this, keeping in mind the sub-Saharan Africa's need for strategies that are affordable and participatory in nature so as to ensure sustainability.

This leads us to an important question: *Is technology an ultimate answer to food insecurity in sub-Saharan Africa?* Africa not only has failed to keep up with the developed world by adopting the so-called appropriate technologies, but it has been unable to capitalise on own indigenous technologies. Nevertheless, indigenous technologies, which are actually traditional practices on their own cannot effectively address the food requirements of Africa's growing population, whereas on the other hand modern technologies are just not sustainable, because they have ignored traditional practices and serve mainly the urban population, which represents a very small segment of the population. Over 70 per cent of Africa's population is ruralbased and is far removed from the modern technologies, which have been transferred into the African continent. Unfortunately for Africa, the world has become a global market and countries can no longer operate on isolation, therefore the countries in Africa have to keep up with globalisation. Lately, biotechnology has been the bone of contention between the impoverished Africa and the developed world. Hunger has been politicised and governments in Africa have been under pressure to accept genetically engineered foods as food aid, whereas the underlying health risks are not really known. There is need for proper information regarding the products so that with the information at hand, then the ultimate responsibility and decision regarding the acceptance and distribution of genetically modified foods as food aid will rest with the governments concerned.

Nevertheless, Africa's food system is in transition, a very slow transition. It is important to note that we have on one hand our traditional practices, which on their own cannot effectively address the food requirements of Africa's growing population. On the other hand are the modern technologies, which are just not sustainable, because they have ignored traditional practices. There is need to marry the two.

According to Bunders et al (1996) sustainable food security can be achieved only through a higher productivity per units of land, water, energy and time. All this calls for the integration of the best in traditional ecological wisdom and prudence and modern technologies. Therefore, indigenous biotechnology is an urgent necessity if the concept of sustainable agriculture is to become a reality at the field level.

Conclusion

Nutrition is an excellent investment. Improved nutrition empowers people and it empowers communities. In doing so it fuels the development process and leads to poverty reduction (Haddad, 2002). Ending hunger in the world is a cause, which would provide the driving force for an equally effective international movement driven by civil society. Failure to address the problem of hunger frustrates the achievement of the now well accepted goal of poverty alleviation, yet eradicating hunger has not been formally singled out either in the international development goals, or as an objective for poverty reduction strategy processes (PRSP) which are now being implemented in most sub-Saharan African countries. It is worth noting that alleviation of hunger is not just a moral imperative, but also the fulfilment of a fundamental human right. The long-term answer to the problems of poverty, food insecurity and malnutrition according to FAO (1997) is sustainable economic growth coupled with equitable distribution of that growth throughout society.

The UN provides the lead and the other intergovernmental organisations have the responsibility of following this lead by providing the framework in which to operate, whereas the various states must provide a stable political arena as well as the will and commitment towards the achievement of the goals. Meanwhile, civil society does have an important role in this by putting to task these organisations to ensure that goals are achieved. Let us take nutrition seriously as an input to development, through adopting strategies that integrate agriculture, food and nutrition interventions. This is a challenge to the sub-Saharan countries to see nutrition and nutrition improvement as a contribution to productivity, poverty reduction and national development. There is need for governments to invest more on community-based intervention programmes. As a point of caution, in as much as emphasis is being placed here on community based food, nutrition and agricultural projects, it is worth noting that they are essential strategies which are necessary but not adequate in improving long-term outlook on sustainable agriculture and, eventually, development. There is need for integration of other efforts in a synchronised manner to tackle fundamental causes of poverty, food insecurity and malnutrition.

References

ACC/SCN. 2001. "What Works? A Review of the Efficacy and Effectiveness Nutrition Interventions," Allen, L. H. and S. R. Gillespie. ACC/SCN: Geneva Collaboration with the Asian Development Bank, Manila.

——— 2000. *4th Report on The World Nutrition Situation*. Geneva: ACC/SCN in Collaboration with IFPRI.

——— 1998. Papers from the ACC/SCN 25th Session Symposium, Oslo, Norway March 30 and April 1, 1998. *Symposium Report Nutrition Policy Paper* 17. November 1998.

——— 1997. "Nutrition and Poverty Papers from the ACC/SCN Symposium", *Report Nutrition Policy Paper* 16. November 1997. Geneva.

Bunders, J., B. Haverkort, and W. Hiemstra. 1996. *Biotechnology*: *Building on Farmers Knowledge.* Macmillan Education Ltd. London.

FAO. 1997. *Agriculture, Food and Nutrition for Africa: A Resource Book for Teachers of Agriculture.* FAO Publishing Management Group Rome.

Haddad L. "Nutrition and Poverty" *Nutrition: A Foundation for Development.* Geneva ACC/SCN. 2002.

SCN NEWS. 2002. "School-age Children Their Health and Nutrition." *United Nations System* 25. December 2002. Geneva.

——— 2001. "Civil Society and the UN System: Debate in the Food Security Arena." *United Nations System's Forum on Nutrition* 25. December 2001. Geneva.

——— 1997. "A Periodic Review of Developments in International Nutrition." *Compiled from Information Available to the ACC/SCN* 15. December 1997.

Maxwell, S. and T. R. Frankenberger. 1992. *Household Food Security: Concepts, Indicators, Measurements.* A Technical Review UNICEF Publication.

Moser C. O. N. 1996. *Confronting Crisis: A Comparative Study of Household Responses to Poverty and Vulnerability in Four Poor Urban Communities*. The World Bank.

Ogunrinade, A.., R. Oniang'o, and J. May. 1999. *Not by Bread Alone: Food Security and Governance in Africa*. Toda Institute for Global Peace and Policy Research ABC Press South Africa.

UNICEF. 1993. *The State of the World's Children*. Oxford University Press.

UNEP. 2002. *Assessment of Progress on Sustainable Development in Africa since Rio (1992)*. United Nations Environment Programme, Regional Office for Africa, Nairobi, Kenya.

THE ROLE OF HOME GARDENS IN IMPROVING HEALTH, NUTRITION AND FOOD SECURITY IN SUB-SAHARAN AFRICA

A. Musotsi,

M. O. A. Onyango,

A. Sigot

Abstract

Food security has been an issue of concern in sub-Saharan Africa, with a rapid population growth averaging 2.8% per annum and a decreasing agricultural production resulting into a decreasing availability of food and malnutrition. Food production in home gardens can significantly contribute to food security and good nutrition at the household level. However, this has not been fully exploited. A study was conducted in Western Kenya between January 2002 and June 2003 with the following three objectives: to determine methods of food acquisition by rural households in Western Kenya; to determine influence of home gardening on household food security and to collect and standardise recipes of African Leafy Vegetables in Western Kenya. A descriptive survey design was used in the study, simple random and purposive sampling were employed to obtain 100 households, 20 key informants and 40 members of local self help groups, structured questionnaire, interview schedule and panel group discussions were used to collect data. Results indicated that the three methods used for food acquisition in the study area were agricultural production from home gardens, market purchases and food gathering. Home gardening was found to influence household food security with respect to the land size and number of livestock. Five recipes of African Leafy Vegetables were collected and standardised. The results obtained will contribute to efforts to improve the role of home gardens in household food nutrition and security in sub-Saharan Africa.

Key words: food security; nutrition; sustainable home gardening; African leafy vegetables recipe collection and standardisation.

Introduction

Background

Food security has been an issue of concern in many parts of the developing world. While developed countries exhibit high technological advancement in food production, gardening remains the most important method of food production for majority of people in developing countries. The economies of most sub-Saharan African countries are dominated by agriculture and related activities, representing about 30 - 35% Gross National Product (Werblow, 1996). However, agricultural production has slowed down to 1.3% in Africa since 1980. A population growth rate averaging 2.8% per annum over the last three years has meant an increasing food gap and decreasing availability of food supply (Werblow, 1996) leading to chronic food insecurity in all sub-Saharan African countries.

Food insecurity has been identified as the prime cause of malnutrition in many sub-Saharan households. Hunger continues to be widespread and the right of everyone to adequate food is extensively violated (this right had been proclaimed in 1948 by the General Assembly, through the adoption of the Universal Declaration of Human Rights. It was reaffirmed in 1996 by the World Food Summit and Food and Agriculture Organisation of the United Nations). Eide (1999) argues that while food supply has substantially increased globally, constraints on access to food and continuing inadequacy of incomes at household and national level to purchase food, instability of supply and demand, as well as natural and man-made disasters, have prevented basic food needs from being met. As food becomes scarce at household level, many families have to go without food for several days.

Food production in small quantities, mainly from well managed home gardens, significantly contributes to food security and good nutrition at the household level. They can supplement food needs during the lean season and generate income. Leafy and fruit vegetables are the major crops found in home gardens and these form a significant part of the diets of many rural communities. African Leafy Vegetable consumption and knowledge on ways of preparing the vegetables have declined since colonial days, these being brought about mainly by interaction with other cultures. In the recent past, government and non-governmental organisations have made deliberate efforts to increase the production and consumption of African Leafy Vegetables.

Some of the important African leafy vegetables, which have been identified, include leaf amaranths (*Amaranthus blitum/cruentus*), slenderleaf (*Crotalaria brevidens/ochroleuca*), cowpeas (*Vigna unguiculata)* African nightshades (*Solanum scabrum/villosum*), spiderplant (*Cleome gynandra*), Jute mallow (*Corchorus olitorius*), African kale (*Brassica carinata*), Pumpkin leaves (*Cucurbita moschata/maxima*) (Chweya, 1985; Schippers, 2000; Onyango, 2002*a*)

Limited research also indicates that these vegetables have high nutritive value. They contain high levels of minerals especially calcium, iron and phosphorus, vitamins and proteins (Chweya, 1985; Imbamba, 1973). In most cases the mineral and vitamin contents supersedes that found in popular exotic vegetables like cabbage, lettuce and spinach (Imbamba, 1973; Adams and Richardson, 1977; Gomez, 1981, 1982; Imungi and Porter, 1983). African leafy vegetables are therefore a valuable source of nutrition in rural areas where they contribute substantially to protein, mineral and vitamin intake (Mnzava, 1997). A daily consumption of 100g of fresh leaf of most of the African leafy vegetables would provide more than 100% of the recommended daily intakes of the vitamins and minerals (Chweya, 1985).

African leafy vegetables can play an important part in food security, especially during periods of scarcity, they can be preserved and used to fill in the relish gap (Mnzava, 1997) and can be important in household food security. In addition, most of these vegetables have been reported to have medicinal properties (Olembo *et al,* 1995) Spiderplant and jute mallow have been reported to aid constipation and spiderplant can facilitate birth. African nightshades and slenderleaf have been reported to cure stomach-ache and ulcers (Olembo *et al,* 1995; Schippers, 2000).

Some agronomic advantages of African vegetables include short growth period (Schippers, 2000) seed production under tropical conditions (Onyango, 2002*b*) as opposed to the temperate vegetables, and can withstand stress (Mwai et al, 2002). The ability for these vegetables to grow fast and become harvestable within a short period makes them useful in sustaining nutrition intervention programs. Being accessible to the low-income groups, they offer opportunity of providing affordable nutrition and broaden the food base. Socio-economic studies carried out in West and East Africa indicates that African Leafy Vegetables provides employment and enables the poorest people to earn a living by growing and trading in these vegetables (Schippers, 2000; Onyango, 2002*a*).

Statement of the Problem

Access to nutritionally adequate diets in sub-Saharan Africa remains unattainable because food production strategies have not exploited the potential of home gardens to ensure food security, (FAO, 1997). Africa is endowed with a high agricultural biodiversity, yet prevalence of hunger and general malnutrition leading to underweight, marasmus and kwashiorkor and even death are frequently reported. Home gardening presents itself as a way of directly accessing a diversity of nutritionally rich foods (FAO, 2001). A variety of fresh garden foods on a daily basis significantly improves the quantity and quality of nutrients available to the family (Muelhoff, 2001). There is a need for looking at recipes of foods, particularly African vegetables to maintain the nutritive value since diets are also affected by preparation methods.

Objectives of the study

Broad Objective (Goal)
To study the role of home gardens in improving the health, nutrition and food security of the people in sub-Saharan Africa.

Specific Objectives

- To determine methods of food acquisition by rural households in Western Kenya.
- To determine influence of home gardening on household food security.
- To collect and standardise recipes of African Leafy Vegetables in Western Kenya.

Research Methods

Time and Area of Study
This study was carried out in Butere-Mumias district of Western Province, Kenya between January 2002 and June 2003. It covers a total area of 939.3sq. km, lying between longitudes 34° 21' East and 34° 41' East and latitudes 0° 15' North and 0° 29' North of the Equator. The district has varying topography. The lowest part stands at about 1,240 m above sea level and the highest at 1,641 m above sea level. The district has a variety of soils resulting from different underlying rocks. The district is also dissected by a number of streams and rivers giving it ample surface water resources. This combined with good climate and underlying rock often gives a high potential for agriculture. In addition, the district has high rainfall ranging from 1,597 - 2, 873 mm per annum. The mean annual temperatures are 29° C, allowing for crop development year round. The population density of the district is 572 persons per sq. km, which is relatively high for an agricultural society.

Research design, data collection and analysis
Descriptive survey design was used and the target population was rural households. The sample consisted of three categories of people and the first category was obtained through simple random sampling where 100 households were selected.

The second and third category employed purposive sampling to obtain 20 key informants from agricultural institutions and 40 others from local self help groups respectively. The latter sample was to help document recipes for African leafy vegetables.

Data was collected using a self-administered questionnaire for 20 key informants and an interview schedule and observation checklist for the households. Panel group discussions and an interview schedule were used to obtain information from the 40 key informants for the recipe collection study. Qualitative and quantitative data analysis was done, where descriptive statistics, frequency distribution and percentages were applied to organise data quantitatively. Correlation's coefficient was used to show relationships between selected variables. The recipes collected were standardised in the foods and nutrition laboratory of the School of Family, Consumer Sciences & Technology, Maseno University - Kenya.

Measurement of variables

The following were indicators used to measure the 2 main variables of the study:

Home gardening: this was measured using:

- Size of land- this refers to the size of land for agricultural production.
- Food production activities- this refers to crops and livestock produced by a household.

Food security- this was measured using:

- Food stock –this refers to the amount of food in store for future use.
- Number of meals eaten daily –this refers to the average number of meals eaten daily by a household.

Standardisation of recipes

The standardisation process of the recipes was done in the foods laboratory at Maseno University. This involved the weighing and measuring of the ingredients within the collected recipes to obtain acceptable measurements and duration of cooking for specific recipes.

Results and discussion

Methods of food acquisition by households in Western Kenya

Results revealed that there were three main sources of food for households in Western Kenya. First, agricultural production was relied on by 100% households. Table 1 shows absolute values obtained of the kinds of food crops grown in order of priority. Maize was given priority because it contributes significantly to the staple food, i.e. *ugali*. *Ugali* is made from plain maize meal or a mixture of maize and/or millet, sorghum and cassava. This finding is consistent with the Butere-Mumias District Development Plan (2001-2006) data, which showed that maize was a priority crop. Vegetables, though not taking the lead in priority are grown by 100% households. This shows how important they are in the diet of these households. Yet, in the statistical picture of food production in western Kenya by the Butere-Mumias District Development Plan, vegetables did not feature. In addition to food crops, 99% of households in Western Kenya also rear livestock as shown in table 2.

Table 3 shows that 56% of the households have land sizes of 2 acres or less and 54% of the households indicated that their size of land was insufficient to produce food that can meet the year round household food requirement. This observation is in agreement with the District Poverty Assessment Report (2000), which showed that food poverty was caused by, among others, high population leading to land fragmentation and causes these households to experience food scarcity.

Table 1: Crops grown in order of priority:

Crops priorities	Maize	Millet	Sorghum	Legumes	Nuts	Root crops	Bananas	Vege-tables
1	92	-	-	1	-	-	3	5
2	2	4	4	68	2	6	4	9
3	5	3	6	14	17	16	10	29
4	-	3	3	8	11	27	19	20
5	-	-	-	4	9	15	27	23
6	-	3	1	1	5	4	12	9
7	-	1	2	-	-	1	3	2
8	-	1	-	-	1	-	2	1
9	-	-	-	-	-	-	-	1
N/A	1	85	84	4	55	31	20	1
TOTAL	100	100	100	100	100	100	100	100

Table 2: Number of livestock

	Frequency	Percent	Valid percent	Cumulative percent
Few	83	83.0	83.0	83.0
Moderate	16	16.0	16.0	99.0
None	1	1.0	1.0	100.0

Table 3: Size of land for agricultural production

	Frequency	Percent	Valid percent	Cumulative percent
0-1 acre	31	31.0	31.0	31.0
1.1-2	25	25.0	25.0	56.0
2.1-3	30	30.0	30.0	86.0
3.1-4	6	6.0	6.0	92.0
4.1-5	2	2.0	2.0	94.0
5.1 and above	6	6.0	6.0	100.0

Market purchases of food is another source that was employed especially in times of scarcity, and to supplement that which is produced within the household. Results showed a high rating for this method of food acquisition, with 88% households employing it. Due to high levels of poverty among rural households in western Kenya many households go hungry, because they cannot afford to purchase food from the markets.

Lastly, gathering as a method of food acquisition was employed by a small percentage of households (12%). This method is becoming less popular due to loss of genetic resource and lack of fallow and forestland, where wild plants, insects and animals are found. The few species gathered include wild insects, fruits and vegetables and birds. Whereas this method is still significant today, there is need to domesticate some of these species so as to prevent their extinction.

The influence of home gardening on household food security

Size of land and food stock

The correlation between size of land and food stock was significant ($p \leq 0.001$) with a correlation coefficient of 0.336 as shown in table 4. This implies that about 34% of changes in food stock were caused by a corresponding change in size of land and when the size of land increases food stock increases and vice versa. Households with larger pieces of land had more food stock as compared to those who had smaller pieces of land. Land size still affects a household's food stock in Butere division probably because of poor land use practices as noted by the District Poverty Assessment Report (2000). Over-dependence on cash crops, mainly sugarcane has adversely affected food production. Also households have neither intensified nor diversified food production on the small

Table 4: Correlation of Size of land and food stock

		Size of land	Food stock
Size of land	Pearson Correlation	1.000	.336**
	Sig. (2-tailed)	.	.001
	N	100	100
Food stock	Pearson Correlation	.336**	1.000
	Sig. (2-tailed)	.001	.
	N	100	100

**Correlation is significant at the 0.001 level (2-tailed).

sizes of land available hence production is still very low. Households should, therefore, increase the land size for food production and reduce on land for cash crop farming.

Size of land and number of meals

There was no significant (p>0.05) correlation between size of land and number of meals. A correlation coefficient of 0.133 was obtained indicating that only 13% of changes in the number of meals resulted from changes in size of land. The size of land that a household owns, therefore, had little influence on the number of meals eaten in the households.

Table 5: Correlation of size of land and number of meals eaten daily

		Size of land	No. of meals eaten daily
Size of land	Pearson Correlation	1.000	.133
	Sig. (2-tailed)	.100	.188
	N		100
No. of meals eaten daily	Pearson Correlation	.133	1.000
	Sig. (2-tailed)	.188	.
	N	100	100

This observation could be explained by the fact that methods of obtaining food like buying and gathering may be employed. Some households bridge the gap between food scarcity and number of meals by buying or gathering thus being able to avoid changes in number of meals as a result of small land size. Rather than rely on buying food, which is an expensive undertaking, households should be encouraged to produce enough from their land.

Home garden crops and food stock

The correlation between home garden crops and food stock was not significant (p>0.05) with a correlation coefficient of 0.117, indicating that only 12% of changes in food stock were as a result of changes in home garden crops.

Table 6: Correlation of home garden crops and food stock

		Home garden crops	Food stock
Home garden crops	Pearson Correlation	1.000	.117
	Sig. (2-tailed)	.	.248
	N	100	100
Food stock	Pearson Correlation	.117	1.000
	Sig. (2-tailed)	.248	.
	N	100	100

These findings imply that in western Kenya, home gardening is not an established practice. This agrees with the data given by Butere-Mumias District Development Plan (2001-2006), that the main food crops produced are not home home garden crops and number of meals eaten daily garden crops. Also, FAO (2001) noted that home gardening is still a neglected practice in many communities. Despite the capacity of home garden crops to affect food stock, households in western Kenya do not experience this.

Home garden crops and number of meals eaten daily

There was no significant (p>0.05) relationship between home garden crops and number of meals eaten daily the correlation coefficient was 0. The value implies that changes in home garden crops do not at all affect changes in number of meals eaten daily. A possible explanation for this observation is that many households have resorted to selling their foodstuffs in order to meet other basic needs. Data from key informants elicited 75% responses to the affirmative for households who sell their farm produce. Most of what is sold are home garden crops leading to insignificant effect to the number of meals eaten daily. Households should be encouraged to locally consume rather than sell their produce.

Table 7: Correlation of home garden crops and number of meals eaten daily

		Home garden crops	No. of meals eaten daily
Home garden crops	Pearson correlation	1.00	.000
	Sig. (2-tailed)	.	1.000
	N	100	100
No. of meals eaten daily	Pearson correlation	.000	1.000
	Sig. (2-tailed)	1.000	.
	N	100	100

Number of livestock and food stock

The correlation of number of livestock and food stock was significant ($p \leq 0.05$) with a correlation coefficient of 0.211. This implies that 21% of changes in foodstock could be accounted for by changes in the number of livestock. Increase in the number of livestock in the households leads to a corresponding increase in foodstock and vice versa.

Table 8: Correlation of number of livestock and food stock

		No. of livestock	Food stock
No. of livestock	Pearson correlation	1.000	.211*
	Sig. (2-tailed)	.	.035
	N	100	100
Food stock	Pearson correlation	.211*	1.000
	Sig. (2-tailed)	.035	.
	N	100	100

This could be due to the fact that livestock waste is used to produce farmyard manure, which is used in gardens to improve yields resulting in more food stock. This concurs with the reports of FAO (2001) and Muelhoff (2001) who suggested that the home garden could take advantage of the interdependedness in order to reduce cost of maintenance and to produce more food. Promotion of livestock keeping should, therefore, be done so that households have a complete entity of a home garden.

Number of livestock and number of meals eaten daily

The correlation between number of livestock and number of meals eaten daily was insignificant ($p > 0.05$) with a correlation coefficient of only 0.064, showing that only 6% of changes in number of meals occurred due to changes in livestock.

Table 9: Correlation of number of livestock and number of meals eaten daily

		No. of livestock	No. of meals eaten daily
No. of livestock	Pearson correlation	1.000	.064
	Sig. (2-tailed)	.	.527
	N	100	100
No. of meals eaten daily	Pearson correlation	.064	1.000
	Sig. (2-tailed)	.527	.
	N	100	100

A change in the number of livestock appears not to affect the number of meals eaten daily, probably because livestock as a source of food is utilised occasionally. It is not common practice for households to slaughter livestock since their monetary value is high. Some households may also find it hard to consume products such as eggs hence they sell them. Milk as the most commonly consumed product is incorporated in tea, usually accompanying a carbohydrate. Thus if a household takes black tea instead of white tea, it will not affect the number of times meals are taken but the quality.

Collection and standardisation of recipes for African leafy vegetables

The five recipes collected and standardised during the study included *omusundulu, atwaka,* amaranths mixed with spiderplant, *lipooka* and *eshikangulu* as shown in table 10. It was observed that most of the recipes were made of more than one leafy vegetable and the combinations ranged from one to five and the combination of two vegetables was the most common. In most of them other additions were incorporated especially milk or milk products. The mixing of the different types of vegetables and addition of milk was aimed at improving the palatability and the nutritional and medicinal qualities of the vegetables (Schippers, 2000). The bitter types were mixed with the mild tasting ones and the tough textured with the soft textured species.

Recommendations and conclusions

Recommendations

- Stakeholders and households need technical information on effective home gardening so that their potentials in household food security can be realised.

- Households practising home gardens should be advised not to sell their produce till the household's food requirement is met.
- Recipes collected and standardised should be well documented and made available to the households in sub-Saharan Africa.

Suggestions for further research

More research is needed on:

- The nutritive value of home garden foods particularly African leafy vegetables (ALV's) before and after cooking.
- Gender roles in home gardening.
- Preservation techniques for home garden foods that could be adopted by the rural households.

Conclusions

The three methods of food acquisition by households in the study area were agricultural production from the home gardens, market purchases of food to supplement the production and food gathering by a small percentage (12%) of the households.

Home gardening influences household food security with respect to the land sizes and number of livestock but not with home garden crops. Home garden crops and number of livestock have no significant influence on the number of meals eaten.

The five recipes of African leafy vegetables collected and standardised were mostly mixtures of two or more vegetables and this improved palatability, nutritional and medicinal value of these vegetables.

Acknowledgements

We wish to acknowledge the following for their support in enabling us successfully conduct the research: The International Plant Genetic Resource Institute (IPGRI); Prof. Ruth Oniang'o and the Rural Outreach Program (ROP) and Maseno University, Department of Home Science and Technology.

Table 10: Five recipes

VEGETABE	RATIO	METHOD OF PEPARATION	VARIATION
1. Omusundulu Leaf amaranth Slenderleaf Spiderplant African night-shades Pumpkin leaves Watr Salt Milk	 250g (1 large bunch) 1000g (4 large bunches) 500g (2 large bunches) 500g (2 large bunches) 1000g (4 large bunches) 1000 c.c.(4 cups) to taste 450 c.c. (2 cups)	Prepare the leaves. Wash the leaves. Cut the pumpkin leaves. Heat water and add vegetables. Cover vegetables. Boil vegetables for about 45 minutes. Add salt and milk. Cook for another 10 minutes. Vegetables are ready to serve.	Omit pumpkin leaves.
2. Atwaka Leaf amaranth Cowpea leaves African kale Water Salt Milk	 250g (1 bunch) 1000g (4 bunches) 1000g (4 bunches) 1000c.c. (4 cups) to taste 225 c.c. (1 cup)	Pluck the vegetables. Wash vegetables separately. Remove stalks from kale. Cut the kale. Mix all the vegetables. Boil water in a pot/ pan. Add vegetables and cover. Boil for about 30 minutes. Add salt and milk Cook for another 10 minutes. Vegetables ready to serve.	
3. Amaranth mixed with spider plant Leaf amaranth Spider plant Water Salt (optional) Sour milk/ ghee	250g (1 large bunch) 250g (1 large bunch) 675 c.c. (3 cups) 225c.c. (1cup)/ 1 tablespoon	Pluck the leaves. Wash and drain the vegetables. Heat water in a pot/pan. Add vegetables to the water. Cover the vegetables with banana leaves, then a lid. Simmer slowly for about 40 minutes. Remove egetables from the fire and turn them. Mash the vegetables. Add sour milk/ghee and cook for another 5 minutes. Vegetables ready to serve.	Substitute for spider plant: Black nightshade

4. Lipooka Cowpea leaves Leaf amaranth Salt Water	 200g (1 bunch) 200g (1 bunch) to taste 675 c.c. (3 cups)	 Prepare the leaves. Wash the leaves. Heat water in a pot/pan. Mix vegetables with salt on a tray. Add vegetables to the boiling water. Cover with banana leaves then lid. Boil for 25-30 minutes. Vegetables are ready to serve.	
5. Eshikangulu Cowpea leaves Decanted salt Water Salt Milk/ghee	 200g (1 bunch) 56 c.c. (¼ cup) 225 c.c. (1 cup) to taste 115 c.c. (½ cup)/1 teaspoon	 Pluck the vegetables while removing the fibrous parts. Wash and drain. Heat water and decanted salt. Add the vegetables and cover with a lid. Simmer for about 20 minutes. Add fresh milk/ ghee. Cook for another 5 minutes. Vegetables are ready to serve.	 Crotalaria plain can also be prepared in the same way. Fry with cooking fat, onion and tomato.

Table 10: Recipe Collection and Standardisation

References

ACC/SCN Symposium Report. 1997. "Nutrition and Poverty. " Nutrition Policy Paper 16, 69-76. Papers from the 24th Session Symposium. Kathmandu, March 1997. United Nations.

Adams, C. A. and M. Richardson.. 1977. "Nutritive Value of Foods." USDA, *Home and Garden Bulletin* 72.

Bragdon S. 2000. "An Overview of International Instruments of Relevance to Plant Genetic Resources." *International Obligations and National Objectives: Option for Plant Variety Protection in the new Millennium.* Report on the Regional Workshop in South Africa on the Implementation of Article 37.3(b) of the TRIPS Agreement - 22-25 March 1999; Harare, Zimbabwe. Rome, IPGRI.

Butere-Mumias District. 1999. "The State of Poverty and Strategies for Intervention." Seminar Report. Butere Boys High School, 28-30 April 1999.

Chweya, J. A. 1985. "Identification and Nutritional Importance of Indigenous Green Leaf Vegetables in Kenya." *Ninth African Symposium on Horticultural Crops* 99-108. July 1983. G. E. Tidbury and H. D. Tidal, eds. Wageningen, Netherlands. ISHS. E.

Eide A. 1999 "Globalisation, Universalisation and the Human Right to Adequate Food" *Not by Bread Alone. Food Security and Governance in Africa.* A. Ogunrinade, R. Oniang'o, and May J.Toda, eds. Institute for Global Peace and Policy Research, South Africa.

FAO. 1987. *Promoting Underexploited Food Plants in Africa. A Brief for Policy Makers.* FAO. Rome.

——— 1997. *Agriculture, Food and Nutrition for Africa. A Resource Book for Teachers of Agriculture.* FAO. Rome.

——— 2001. *Improving Nutrition through Home Gardening. A Training Package for Preparing Field Workers in Africa.* FAO. Rome.

Gall D. M, W. R. Borg, and J. P. Gall. 1996. *Educational Research. An Introduction.* 6th ed., Longman Publishers. New York, USA.

Gomez, M. I. 1981. Carotene Content of some Green Leaf Vegetables of Kenya and Effects of Dehydration and Storage on Carotene Retention. *Journal of Plant and Food* 3:231-244.

——— 1982. The Evaluation of Fruit and Vegetable Resources in the Machakos District in Relation to Seasonal Deficits and Micronutrient Deficiencies. *Technical Report IDRC* 39.

Grinnell R. M. 1993. *Social Work Research and Evaluation,* 4th ed. F. E. Peacock Publishers, Inc. Illinois.

Imbamba, S. K. 1973. "Leaf Protein Content of some Kenya Vegetables." *East African Agriculture and Forestry Journal* 38:246-251.

Imungi, J. K. and N. N. Porter. 1983. "Nutrient Contents of Raw and Cooked Cowpea Leaves." *Journal of Food Science* 48:1252-1254.

Kujeke R. 1999. "Developing New Strategies for Food Security in the SADC Region." *Not by Bread Alone. Food Security and Governance in Africa.* A. Ogunrinade, R. Oniang'o, and J. May, eds. Toda Institute for Global Peace and Policy Research. South Africa.

Latham M. and Beaudry M. (1999): "The Impact of Transnational Corporations on Food-Consumption and Nutrition in Africa." *Not by Bread Alone. Food Security and Governance in Africa.* Ogunrinade A., R. Oniang'o, and J. May, eds. Toda Institute for Global Peace and Policy Research. South Africa.

Mason J. B., J. P. Habicht , H. Tabatabai, and V. Valverde. 1984. *Nutritional Surveillance. WHO.* Geneva.

Mason, J. E and J. W. Bramble. 1997. *Research in Education and the Behavioural Sciences.* Concepts and Methods. Brown and Benchmark Publishers. Mexico City.

Maundu W. P, W. G. Ngugi, and H. S. C. Kabuye. 1999. *Traditional Food Plants of Kenya.* Resource Centre for Indigenous Knowledge (KENRIK). Nairobi, Kenya.

Mnzava, N. A. 1997. "Nutritional Values of Exotic and Indigenous Vegetables."
Proceedings of a Workshop on African Indigenous Vegetables. pp. 70-75. Limbe, Cameroon. Jan. 1997. R. R. Schippers and L. Budd, (eds). Chatham, UK: IPGRI/Natural Resources Institute.

Maxwell S. and R. T. Frankenberger. 1992. "Household Food Security: Concepts, Indicators, Measurements." *A Technical Review.* New York, UNICEF Rome, IFAD.

Midmore, D. J., V. Ninez, and R. Venkataraman, 1991. "Household Gardening Projects in Asia: Past Experience and Future Direction." *Food, Nutrition and Agriculture: Strategies to Combat Micronutrient Deficiencies.* FAO, 1993. Rome.

Mpoyi, K. and S. J. Paulus. 1997. "Promoting Home Gardening in Kinshasa, Zaire." *Traditional African Vegetables.* Proceedings of the IPGRI International Workshop of Genetic Resources of Traditional Vegetables in Africa: Conservation and Use. 29-31 August 1995, ICRAF – HQ, Nairobi, Kenya. FAO, Rome.

Muelhoff, H. 2001. *How Home Gardens can Significantly Contribute to Good Nutrition Year Round.* FAO, Rome.

Mugenda, M. O and G. A. Mugenda. 1999. "Research Methods." *Quantitative and Qualitative Approaches*. ACTS Press. Nairobi.

Mwai, G. N., J. C. Onyango, and M.O.A. Onyango. 2002. "Potential Salinity Resistance in Spiderplant (*Cleome gynandra*)." A Paper Presented to the 2nd Horticulture Seminar on Sustainable Horticultural Production in the Tropics, August 6th to 9th 2002. Jomo Kenyatta University of Agriculture and Technology, JKUAT. Juja, Kenya.

Okigbo B. N. (1994): "Conservation and Use of Plant Germplasm in African Traditional-agriculture and Land Use Systems." *Safeguarding the Genetic Basis of Africa's Traditional Crops*. Proceedings of a Technical Centre for Agricultural and Rural Cooperation. CTA/IPGRI/KARI. United Nations Environment Program (UNEP). 5-9 October 1992, Nairobi, Kenya. Netherlands, CTA. Rome, IPGRI.

Olembo K. N, S. S. Fedha, and S. E. Ngaira S. E. 1995. *Medicinal and Agricultural Plants of Ikolomani Division, Kakamega District.* Development Partners, Nairobi.

Onyango, M.O.A. 2002a. "Market Survey of African Indigenous Vegetables in Western Kenya." *Proceedings of the Horticulture Seminar on Sustainable Horticultural Production in the Tropics.* August 6th to 9th 2002. Jomo Kenyatta University of Agriculture and Technology, KUAT. Juja, Kenya. J. M. Wesonga, J. M. T. Losenge, G. K. Ndung'u, K. Ngamau, F. K. Ombwara, S. G. Agong, A. Fricke, B. Hau, and H. Stützel, eds.

—— 2002b. "African Indigenous Vegetables-Opportunities and Constraints." *Proceedings of the Horticulture Seminar on Sustainable Horticultural Production in the Tropics.* October3rd to 6th 2001. pp 81-91. Jomo Kenyatta University of Agriculture and Technology. JKUAT. Juja, Kenya. J. M. Wesonga, T. Losenge, G. K. Ndung'u, K. Ngamau, F. K. Ombwara, S. G. Agong, A. Fricke, B. Hau and H. Stützel, eds. ISBN 9966-923-03-9.

Oniang'o R. K. 2001. "Enhancing People's Nutritional Status through Revitalisation of Agriculture and Related Activities in African." *Journal of Food and Nutritional Sciences* 1(1): 43-49. AJFNS. August 2001.

Pinstrup-Andersen, P., R. Pandya-Lorch, and S. Babu. 1997. A 2020 Vision for Food, Agriculture and the Environment in Southern Africa."*Not by Bread Alone. Food Security and Governance in Africa.* Ogunrinade A., R. Oniang'o, and J. May, eds. Toda Institute for Global Peace and Policy Research. South Africa.

Poverty Eradication Commission. 2000. "District Poverty Assessment Reports." Republic of Kenya, Office of the President. December 2000. Nairobi.
Republic of Kenya. Ministry of Planning and National Development. *Butere-Mumias District Development Plan 2001-2006.*

Schippers, R. R. 2000. "African Indigenous Vegetables an Overview of the Cultivated Species." Chatham, UK. Natural Resources Institute/ACP-EU Technical Centre for Agricultural and Rural Co-operation.

Sigot A. J. 2001. "Indigenous Food Systems: Creating and Promoting Sustainable Livelihoods." Paper Presented at International Conference on Indigenous Knowledge Systems: African Perspectives. 12th to 15th September 2001. University of Venda for Science and Technology. Thohoyandou.

Smitasiri, S. and G. Attig. 1992. "Encouraging Production and Consumption of Vitamin-A Rich Foods." *Food, Nutrition and Agriculture: Strategies to Combat Micronutrient Deficiencies.* FAO, 1993. Rome.

Werblow U. 1996. "The ACP-EU Courier". March-April edition. *Not by Bread Alone. Food Security and Governance in Africa.* Ogunrinade A., R. Oniang'o and J. May, eds. Toda Institute for Global Peace and Policy Research. South Africa.

Studies in sub-Saharan Africa

Herausgegeben von Support Africa International Inc.
vertreten durch Baldur Pfeiffer und Franz-Theo Gottwald

Band 1 Uche C. Amalu / Franz-Theo Gottwald (eds.): Studies of Sustainable Agriculture and Animal Science in sub-Saharan Africa. 2004.

Band 2 Franz-Theo Gottwald / Susan Keino / Timothy Rotimi Fayeye (eds.): Fostering Subsistence Agriculture, Food Supplies and Health in Sub-Saharan Africa. 2007.

www.peterlang.de